About the author

Bestselling author Liz Harfull is passionate about telling the stories and unearthing the histories of the extraordinary everyday people who make up our communities, especially in rural and regional Australia.

She has written two international prize-winning books about Australian home cooks and show cooking traditions—*The Australian Blue Ribbon Cookbook* and *Tried, Tested and True*—and two books capturing the life stories of rural women, including national bestseller *Women of the Land* and *City Girl, Country Girl*, which was inspired by her mother.

Her most recent book, *The Women Who Changed Country Australia*, celebrated the centenary of the Country Women's Association in New South Wales, where the iconic women's movement started.

OTHER BOOKS BY LIZ HARFULL

Women of the Land
The Australian Blue Ribbon Cookbook
City Girl, Country Girl
Tried, Tested and True
The Women Who Changed Country Australia

A *Farming* LIFE

TALES OF RESILIENCE FROM INSPIRING RURAL WOMEN

LIZ HARFULL

First published in 2023

Allen & Unwin
Cammeraygal Country
83 Alexander Street
Crows Nest NSW 2065
Australia
Phone: (61 2) 8425 0100
Email: info@allenandunwin.com
Web: www.allenandunwin.com

Allen & Unwin acknowledges the Traditional Owners of the Country on which we live and work. We pay our respects to all Aboriginal and Torres Strait Islander Elders, past and present.

A catalogue record for this book is available from the National Library of Australia

ISBN 978 1 76029 111 2

Set in 12/18 pt pt Sabon LT Pro by Post Pre-press Group, Australia
Printed and bound in Australia by the Opus Group

10 9 8 7 6 5 4 3 2 1

The paper in this book is FSC® certified. FSC® promotes environmentally responsible, socially beneficial and economically viable management of the world's forests.

Dedicated to every young girl who dreams of becoming a farmer

Contents

Introduction

There is no denying that the past few years have tested most Australians, no matter where they live. News reports from rural and regional communities have been filled with stories of loss and devastation linked to unrelenting drought, unending floods, unprecedented bushfires, storms and tropical cyclones. Add to that a worldwide pandemic and the financial challenges brought by soaring inflation, trade embargos and labour shortages. It has been a tough time to be a farmer, or someone who relies on farming for their livelihood.

With these challenges has come a growing awareness of the importance of resilience—both for individuals and their communities. A myriad of studies and support programs have been rolled out to investigate what makes one person or place more resilient than another, and to help build resilience so rural communities are better able to cope with climate change and disaster.

A Queensland study in 2007[1] explored the characteristics that rural people believe contribute to individuals being resilient. Unsurprisingly, they included the ability to bounce back or move on, resourcefulness, a positive attitude, willingness to embrace change and adapt, ingenuity and creativity, having a vision for the future, and being prepared to work hard and have a go, even though you might not succeed. Being open to seeking help from others and having a sense of humour were in the mix too, and so was having a strong sense of connection to the land—a recurring theme with so many of the farmers I have written about over the years.

Combined, the women in this book reflect all those characteristics. They haven't always. Pushed to their physical and emotional limits, their resilience has been honed and tested by terrible loss and life-threatening illness, difficult home lives and fractured relationships, as well as natural disasters. In every case, they have picked themselves up and moved on, supported by family and friends, and a strong sense of spiritual connection to the land they farm and the communities in which they live.

Although I selected women from totally different backgrounds and parts of the country, sometimes there were unexpected connections. The most startling was sitting down for a sunset drink on one of Australia's most remote cattle stations and discovering the woman next to me, Catherine, went to the same little one-room school as one of the other women I was writing about, Ruth, who lives almost 2000 kilometres away. Ruth's sister, Mary, and Catherine were best friends at school

1 D.G. Hegney, E. Buikstra, P. Baker, C. Rogers-Clark, S. Pearce, H. Ross, C. King, A. Watson-Luke, 'Individual resilience in rural people: a Queensland study, Australia', *Rural and Remote Health* 7: 620 (online), 2007

and a similar age. Further conversation revealed a past station hand was from a dairy-farming family just down the road from the dairy farmers I chose to write about, and was attending the same agricultural college in Victoria as their daughter.

Another element common to these women is their optimism for the future of Australian agriculture, despite all the challenges. I spent years as a rural journalist and communicator listening to older farmers speak pessimistically about the prospect of smart, well-educated youth returning to the family farm when they had so many other options—professions with reliable incomes and regular hours, and urban lives with access to better services and more diverse leisure pursuits. But the women in this book have a strong sense of hope from encounters with so many young people who are passionate and excited about following in their footsteps. In several cases, daughters and granddaughters are keen to come back to the family farm, inspired by their mothers and grandmothers.

Wanting to give the next generation every encouragement, many of the women are making time to mentor young people, to become involved in organisations that support them or to fight for issues that will make living in the bush more appealing. One is even contemplating a career in federal politics.

In recent years, I have been fascinated to see that when media crews interview someone from the rural sector about the impact of disaster or difficult times, the person fronting the camera has been, more often than not, a woman. And a woman proudly wearing the label 'farmer'. Even ten years ago this was relatively rare. That's not because there were fewer women farming then, it was more that many were reluctant to be in the limelight, or presume to speak on behalf of their industry.

Of course, women have long been part of the farming scene in Australia. From the first years of European settlement, they accompanied their families in pioneering ventures. During the goldrush years of the mid-1800s, more than a few were left to manage properties while the men raced off to make their fortunes. Future generations did the same thing during two world wars, keeping the farms going and sustaining their families, while also marshalling energy to support the war effort through volunteer groups such as the Country Women's Association and Red Cross.

And yet when I set out to write my first book about rural women (*Women of the Land*), and chose to focus on those running their own farms, I was told that I would struggle to find women who had willingly taken on the challenge. No woman would be foolish enough to attempt it on their own unless they were forced by circumstance, and I would certainly have a hard time finding women running commercial farms full-time if there were able-bodied men in their lives willing to do it. I soon discovered that the greatest dilemma wasn't finding women to write about but selecting which stories to tell.

In this latest collection, I have chosen once again to focus on women who are active farmers. How they came to take this path in life varies. In one case, running their own enterprise began as a defiant act because they were told 'women can't farm'. For others, it stemmed from lifelong ambition, and then there are those who took it on willingly after falling for a man from the land.

One dreamt as a child of marrying a 'cowboy' and headed outback where she met and fell in love with a station owner's son. Two were born into it, taking over running family farms after their fathers willingly handed over the reins in their working lifetime, even though, in one case, the next generation included sons. Then there is the family where mother and daughters are working together to run a highly successful, large-scale enterprise, with the encouragement and support of their admiring partners, who know better than to interfere.

In 2012, I was honoured to have veteran *Landline* reporter Pip Courtney launch *Women of the Land* at a rural press club gathering in Adelaide, where she spoke about the changes she had seen over almost twenty years covering rural issues for the popular Australian Broadcasting Corporation (ABC) television program. She had been working at *Landline* for a year when the ABC announced it was sponsoring a new award celebrating the Australian Rural Woman of the Year. At the time, one of the key issues being debated was the legal status of women who worked on farms, and, their 'right' to claim legally, and without embarrassment, the title 'farmer', as opposed to 'farmer's wife'.

Pip recalled putting together a piece about the inaugural winner, Deb Thiele, a grain-grower from the South Australian Mallee. 'It was remarkable the stir it caused when we showed pictures of her driving a massive John Deere tractor. It was assumed it was a set-up for the cameras and that she didn't really drive this big beast, which of course she did,' Pip said.

Back then, Pip reminded people, if a woman said that she worked on the farm, it was assumed by many that she was inside doing the books, and just helping occasionally out on the farm when an extra pair of hands was needed during harvest or shearing or mustering. For insurance purposes, if a woman was injured working on a farm, she was regarded as a 'homemaker', not a 'farm worker'.

The women in this book do not call themselves 'women farmers'. They do not consider themselves a curious novelty. They are farmers, plain and simple, and they don't see what they are doing as anything particularly exceptional or unusual. They are not in this book just because of their gender. Although they are all very modest, they are extremely capable, admired by their neighbours and peers for their abilities. And their stories are particularly inspiring at a time when so many of us are searching for insights into how we can move ahead after some tough years and build our own resilience.

Liz Harfull
January 2023

1

What the Heart Feels

AMBER DRIVER, ELKEDRA STATION, NORTHERN TERRITORY

Amber Driver is in one of her favourite places. Sitting at the controls of her family's single-engine Cessna, she is flying above Elkedra—a remote cattle station in Central Australia. Dirt tracks thread thin red ribbons across the arid landscape below, weaving through rocky ridges, low-lying hills, golden plains of spinifex and scrubby woodland. It's a typical outback scene until the plane banks to approach the station runway. Glinting in the morning sunlight are two long expanses of dark water. And there is the homestead complex surrounded by acres of emerald-green lawn.

Flying into Elkedra is a magical experience. By air, it's a glorious 80 minutes north-east of Alice Springs. By road, it's a torturous, bone-rattling five hours, even when the notorious Sandover Highway has been graded and nothing goes wrong. Wet weather had made the highway impassable when Amber first visited the station more than twenty years ago, so her future father-in-law was sent to fetch her in the plane.

Marriage wasn't on Amber's mind at the time. She was a young jillaroo from New South Wales, living her childhood dream of riding horses and mustering cattle in the Northern Territory. It was proving just as exciting as she had hoped. For the previous eighteen months Amber had been working on a property closer to Alice, soaking up every opportunity to learn about station life and the skills required to survive and manage cattle in the bush.

Then love intervened—love for a lanky station owner's son, for the place his family had called home for three generations, and for the way of life. An instinctive nurturer, Amber likes to ask, 'How does it make your heart feel?' when people confide their troubles. She has had plenty of her own. Drought, intense heat, dust, isolation and long days of hard physical toil have failed to dim her passion for a place that demands stoic resilience, but intense personal grief and a brush with cancer have tested her to the limit. Still carrying the scars, Amber has emerged a fierce advocate for the community that embraced her as a wide-eyed stranger and made her one of their own.

Amber was around seven years old when she first dreamt of moving to the outback and marrying a cowboy. Her daydreams were fuelled by a handsome young man called Scotty, who worked on her parents' farm and was a very good horseman. After Scotty left to jackaroo on cattle stations in western Queensland and the Northern Territory, Amber longed for the summer holidays. He would always come home for Christmas, and ride over on his handsome chestnut horse to visit her parents. 'I remember sitting on the back step of the house, and my anxiety level would be crazy. I'd just have my heart in my mouth, waiting for Scott. I imagined myself being older, marrying him and going to the Territory and living happily ever after.'

Born in Gympie, Amber spent her first years in the small Queensland community of Rainbow Beach, an idyllic location with beautiful beaches and a subtropical climate. Her father, Jerry Killen, had grown up on a large property between Wee Waa and Narrabri in northern New South Wales. Jerry was the youngest of six children in a well-known family with extensive pastoral holdings. He left home determined to make his own way in the world, working in a range of jobs, from jackarooing to flying charter services after he gained a commercial pilot's licence. Amber's mum, Lexie, grew up on a sheep and cattle property near Coonamble, on the central-western plains of New South Wales. She trained as a nurse at the Sydney Children's Hospital, and became a qualified midwife before moving to Narrabri in 1975 to work in the local hospital's maternity ward.

Lexie met her future husband after some determined matchmaking by a well-meaning patient. When she visited the patient at home one day to check how she was getting on with her new baby, the woman excused herself and rang her husband, who worked with Jerry. 'Come home now and bring Jerry for a visit,' she ordered. Lexie was sitting in the lounge room chatting when he walked in. 'And bing! Oh my goodness, that was about it,' she laughs.

Jerry and Lexie married in a simple ceremony and settled at Rainbow Beach, where Jerry and a mate set up one of the first organised tours to nearby Fraser Island, using vehicles they built from the chassis of old Land Rovers. The Killens' first child, Natasha, was born in 1979. Amber came along almost precisely a year later, followed by William in 1984. Then, when Amber was about four, her parents moved back to the Narrabri area because Jerry had the opportunity to purchase a portion of his family's original property, Abbey Green. He ran a mixed farming enterprise there, with beef cattle and irrigated cropping, mostly cotton.

As soon as they were old enough, the children went to Narrabri West Public School, proudly calling themselves Westies to differentiate from students at the town's other schools. An otherwise bright pupil, Amber caused her parents some concern when she refused to read. She suspects now that she did not want to be compared with Natasha. 'My sister had read every book by the time she got to school, but in Year 5 I was still refusing to read, probably because my sister was so good at it. I could read, but it wasn't my thing,' she explains cautiously. Her parents unwittingly made the situation worse when they attempted to entice Amber by giving her a beautiful book with a pink cover and gorgeous illustrations. It was titled *Stories for Nine-year-olds and Other Young Readers*.

She was eleven at the time, and deeply offended. 'I didn't even open it,' Amber says.

When they weren't at school, all three children helped on the farm. Because it was located about 30 kilometres from town, over mostly dirt roads, opportunities to play sport or participate in after-school activities were limited. Amber remembers taking dancing lessons when she was in Year 6, after being told by her parents that she could choose just one thing. She loved dancing, and it saved making the long trip home on the school bus, with Abbey Green one of the last stops on the run. Instead, Lexie came to collect her; by then she was back nursing at the Narrabri hospital.

One night a week, Lexie also drove them to Wee Waa for taekwondo sessions. She waited in the car, until the instructor eventually suggested that she come in and join the class. 'My sister got about halfway through the levels before boarding school,' Amber says. 'Then it was my brother, myself and Mum. We just loved it. We even competed in some tournaments and Mum actually went all the way through and did her black belt.'

Amber looks back at these years with great affection and deep gratitude for the grounding her parents provided, encouraging their children to believe in themselves and work hard for what they wanted, rather than expect things to be given to them. 'It was a very basic life but we were kings because we were loved. I remember at Christmas time it was so exciting to get a present, and it would be a book we needed for school, and a pair of shoes and a towel, which we thought were the best presents ever. I had one doll and Mum sewed all our clothes. Mum and Dad worked hard but they still made an effort to take us away for a week's holiday somewhere, and Dad always made time to teach us things, like pointing out an ant carrying a stick across to his mates. As we got older our eyes would roll. "Yes, Dad, you showed us last summer." It was a very simple life, but we weren't pushed to the side—we were included in what was going on.'

Only twelve months apart in age, Amber and Natasha were very close, although Amber admits that while her more studious sister would rather be inside, she preferred spending time out on the tractor with her father. 'We were like twins. We did everything together,' Amber says.

Then came the day when Natasha turned twelve and was sent away to an Anglican church boarding school in Tamworth. Amber missed her big sister desperately, looking forward with great excitement to the following year when she would be old enough to join her at Calrossy. About two hours from the farm, it was the closest boarding school for girls.

Although Amber was athletic and a capable scholar, attending Calrossy proved a mixed experience. For a start, Natasha had made new friends and wasn't necessarily all that keen to have her younger sister tag along. 'That hit hard with me,' Amber admits. However, she soon made her own friends and embraced the opportunity to play team sports, in part because it provided an opportunity to leave the campus. She signed up for basketball, indoor soccer, touch football and, her favourite, hockey. She even gave water polo a go, until training in an outdoor pool during freezing weather put her off.

Boarders were accommodated on the campus, and usually not allowed out at weekends. To stay in touch with home, girls could write letters or call from the single payphone in the boarding house. Amber wasn't a 'pen-and-paper girl' but Lexie was an avid letter writer. Even better, her correspondence was usually accompanied by homemade Anzac or ginger biscuits.

The older Amber became, the more she struggled with being at the school, although Year 12 was better because she was allowed to go home at weekends. Amber liked learning and did well enough academically, but she disliked the strict rules imposed to keep boarders in line. 'I felt too confined, too much in a box. There were a lot of subjects that I loved but bush kids grow up with so much freedom and responsibility at home, and then you have a lot of conformity at school—for good reason, because they don't want a couple of hundred girls running wild day and night, but I didn't understand that. My life was bigger than the gates of the school so I was pretty happy to get out of there really.'

By the time she left school at the end of 1998, her parents had taken the pragmatic step of selling the farm and moving into Narrabri. For the farm to remain viable meant borrowing money to buy more land at a time when interest rates were high. Given none of the children was interested in taking the property on, it made more sense to sell. Not appreciating the issues behind their decision, Amber wasn't thrilled about it, although she's come to understand it since.

As a teenager, Amber no longer dreamt of marrying Scotty, who she keeps in touch with to this day, but she hadn't given up on the idea of going to the Territory. She was absolutely convinced working on a cattle station was the best way to discover what she wanted to do with the rest of her life. 'And that's what I told myself. I'll give it a crack for twelve months and if it's not what I think it is, then I can figure out what I need to do next.' She wasn't interested in the more conventional avenue of becoming a governess and tutoring children. For Amber, the Territory meant mustering cattle on horseback. 'Horses, cattle work and cowboys—that was going to be my life,' she says.

Her goal firmly in mind, Amber made some strategic decisions during the final years of her formal education. Reckoning it might be useful, she selected an elective to do mechanics at the local TAFE campus, and she chose an art project that involved welding. It had the added bonus of allowing her to return home for a few days because the school didn't have its own welding gear. Her parents no longer had a workshop so she negotiated access to equipment at a local engineering business operated by Len Hall. A schoolfriend also took Amber along to pony camp when she was about fifteen, so she could develop her riding skills. While other girls her age rode in the top troop, Amber found herself with the five- and six-year-olds, learning the basics. 'And it didn't even matter. I had the time of my life. I just remember totally loving it.'

Needing money to fund her big adventure, Amber picked up weekend employment at a Narrabri service station. She also took on gardening and landscaping for various people, and paid work on the farm, eventually saving enough to purchase a second-hand WB Holden ute.

By mid-1999, Amber was ready. The ute was packed and she had a job lined up with Garry and Barb Dann, who owned Milton Park and Amburla stations, out on the Tanami, about 120 kilometres west of Alice Springs. Garry had gone to school with Len Hall, who put them in touch. 'I will be forever grateful to have people all through my life who have invested in my dreams,' Amber says. 'Len was really interested in young people and what their future held, and he often had a yarn to me about life after school. As I was getting closer to being ready to go, he sort of said that his mate had a station and I should give him a ring. I thought that sounded pretty cool. It felt like that was a turning point for me. Dad didn't organise it. It was me. In truth, if it wasn't for Len, I probably wouldn't have got the job, but I made it happen.'

Jerry and Lexie gave their nod of approval after meeting the Danns when they visited Narrabri to see Len. 'I liked them instantly,' says Lexie. 'Barb was a real mother; she talked about her children and caring for them and everything, and I thought, "Right, Barb will look after this young girl and keep her on the straight and narrow for me. She'll be in good hands."'

Amber had a fair amount of bush driving experience despite being only nineteen but she was planning to take some isolated and challenging outback roads, so her parents laid some ground rules. She must stay in a town every night, not just pull up on the side of the road; and she had to call them every evening so they knew where she was and that she was okay.

Amber's first port of call was Brisbane, where she spent a few days with her sister. Then she meandered across central Queensland. Her original intention was to keep heading west on the Plenty Highway, which is mostly gravel with patches of bulldust. When she pulled up at the police station in Boulia to check conditions, the local constabulary nixed the idea. 'They looked at me and they looked at my ute and had a bit of a chuckle, and they said, "Stay on the bitumen", so I went via Mount Isa.'

While the police might have been concerned about a young woman making such a trip on her own, Amber was never scared or worried. She reckoned growing up on the farm and two years' learning mechanics stood her in good stead. Besides, on the first night on the road, she'd stepped outside her motel room in Tambo, taken in the big night sky full of stars, and thought: *This is what I'm talking about, this is perfect!*

Having never been to the Territory, Amber thought it would be flat and red, but as she drew closer to Alice Springs, travelling south on the Stuart Highway, she realised that was far from the case. 'Everything I encountered on that trip was the best. I wish I'd kept a journal and recorded that feeling you get when something is amazing the first time you see it.'

When she got to Milton Park, the Danns were extremely welcoming. Just as Lexie had hoped, they made her feel part of the family. 'They were just the kindest people,' says Amber. 'It's true to say this for so many people in our community up here, but if you are willing to have a go and put in the hard work, people are willing to support you. Garry and Barb did that in spades for me. They taught me so much about everything, even things I thought I knew. When you put things into the perspective of the Territory, you have to relearn a lot of stuff, and they looked after me. I was very fortunate.'

Amber's first weeks at Milton Park were not exactly what she had in mind however. She spent most of them juicing fruit from the station's extensive orchard. 'We did orange juice, then all the lemons, then I had to make cordial,' she says, smiling wryly. The upside was that Amber was sent into

Alice Springs to deliver the orange juice to a list of addresses. One of them belonged to a girl who was just running out the door to play hockey. Her team was a player short and before Amber knew it, she was out on the field. She had been part of a top-level competition in New South Wales where there were strict rules about wearing the correct uniform, but she soon discovered things were more laid-back in Alice. 'People threw some random gear at me. "You'll be fine." And on I went, and it was so good. I loved it.'

Amber ended up playing a couple of games but she couldn't make a regular commitment because of work. 'Everything was golden when I came to the Territory, but that was the start of the list of things that were the compromise. I missed sport, but you just pour your energy into a different area.'

Inexperienced when it came to outback life, Amber initially tended to look at everything through rose-coloured glasses. However, her first year in the Territory coincided with a bad drought in Central Australia, and most of Milton Park resembled a road, it was so bare. When Garry pointed across the flats and told her grass would grow up to her waist in a good season, she found it difficult to believe. *This poor fella, the drought has skewed his memory*, she thought, but, sure enough, when the rains came the feed was magnificent.

Amber soon learnt being a jillaroo involved a lot more than mustering cattle. Most days were packed with all the mundane tasks required to look after the livestock, as well as the station's vehicles and equipment. Determined to be useful, she soaked up every scrap of knowledge and experience thrown her way. 'I think women naturally like to use 30 times more words than blokes, but you quickly discover that asking a bloke a thousand questions a day isn't going to be the best way. So I learnt to limit my questions to the things I couldn't figure out myself or learn from watching,' she says. 'Don't worry, I made lots of mistakes, but because I was willing to learn from my mistakes, people were willing to help.'

Learning how to apply common sense and logic to figure out how to approach a task or to solve a problem on her own proved valuable too. 'You have to work through every potential thing it could be before driving back to the house, because driving back to the house might take half a day. "This isn't starting, why isn't it starting? Have I checked the fuel is turned on?" And you have to remember things you've been taught. There were times when someone said to me, "I've told you this before." You have to be responsible and get it right next time. People don't want to be telling you ten times—once or twice, okay, but not a dozen.'

Amber gained great satisfaction from becoming a competent horsewoman, tying her first truckie's hitch and working out techniques to compensate for being shorter, or not having the same brute strength as the men. Only 165.5 centimetres (5 foot, 5 inches) tall, Amber found alternative ways to get jobs done without asking for help or Garry having to send an extra hand when usually it was a single-person task. 'Things like lifting a truck tyre and putting it back on the carrier under a truck. I might have to jimmy it up with a cheater bar, which might take a minute longer, but I'd still get it done,' she says. 'That's not something I could learn by watching or asking because I can't do it the same way the men do. I had to apply creative thinking.'

The Danns also took Amber under their wing socially, taking her to gatherings where she made new friends. Having grown up in a region where the Killen name was well known and had sometimes influenced how she was treated, she was determined to be accepted on her own merits. 'Hi, I'm Amber,' she'd say on introduction, leaving out her surname and mentioning instead what station she worked on. 'And that was enough,' she says.

Sometimes she made poor choices in her eagerness to enjoy life to the full. Compared with Elkedra, Milton Park was 'in the suburbs' so she had plenty of opportunities to go into town. 'Alice Springs was a wonderful place to be back then and the young people up here were phenomenal—it was pretty cool. I remember desperately wanting to go to a party, and trying to decide how much money I was going to put across the bar, versus how much I needed for fuel for the ute. When I got caught short it was a little bit embarrassing.'

A calendar highlight in Alice was the Christmas party organised every year by a prominent livestock agency. Popular with station people, it attracted everyone from owners and their families to station hands and governesses. Among the guests she met that first summer was John Driver from Elkedra. The same age as Amber, he was a tall 185 centimetres (6 foot, 1 inch), with friendly dark-grey eyes, brown hair and a dry sense of humour.

On her part at least, it wasn't love at first sight but she enjoyed seeing him whenever the opportunity arose. Initially, that wasn't very often. In those days John only came to town three times a year. However, when he did make the trip it was usually for a few days. 'When all the young ones go to town for an event like the Alice Springs Show, you're not just in there for a day. Everyone gets four or five days off and you spend all that time together. You just make the most of it. I felt like we were kids so it wasn't about making life decisions and being serious. It was more about having fun when everyone got together, and things just evolved from there.'

In between visits, John would occasionally ring the Dann household and ask to speak to Amber. 'Are your intentions honourable young man?' Garry would quiz in mock seriousness before handing over the phone.

Then towards the end of 2000, John invited Amber to spend Christmas at Elkedra. 'Just come and see what it's like up here,' he encouraged. Amber planned to go home and spend Christmas with her family, but she agreed to visit first. There was just one snag. When the time came, the roads to Elkedra were impassable because it was so wet. John's solution was to send the station plane, flown by his father, Roy. 'So the first time I met his dad, I was off to a great start,' she jokes. 'John was meant to be concentrating on running the station and I was perhaps someone who might lead his son astray, *and* he'd been sent to town to collect me!'

Amber loved Elkedra at first sight. Like every visitor to this remote place, she was dumbstruck by the homestead's location, on the banks of a deep, wide stretch of the Elkedra River. So much water, filled with raucous birdlife, just steps from the Driver home in the arid centre of Australia where rivers that hold water year-round are not to be found. John had described it beforehand, but smart phones with cameras didn't exist back then and the reality was astonishing.

John's grandfather had been astounded too when he saw Elkedra for the first time decades before. Born in 1902, John Henry Driver grew up in the port town of Albany in Western Australia, where his parents established a small grocery business, and his mother, Mary Ann, was a notable figure in the early women's movement. At the age of twelve, he obtained a scholarship to attend the Perth Modern School, which had recently opened as the state's first government high school. After completing his formal education, he turned down a position as a cadet draftsman with the lands department to join a private surveying practice because he wanted to work outside.

John's first experience of Alice Springs came in 1929, when he started a two-year stint as a surveyor with the North Australia Commission—the federal government authority charged with administering lands, railways and ports in the Territory and planning its development. At the time the town had one hotel, the Stuart Arms; the railway from Adelaide was about 130 kilometres short of being completed, and there was no telephone service, with residents still reliant on the overland telegraph service.

In a brief memoir written for his family, John described Alice as a place in transition. With the coming of the railway and a push to develop the Territory's mining and pastoral industries, it was changing from an isolated shanty town to the thriving service centre Amber encountered seventy years later. The population was still fewer than 500 people when John returned in 1937 to work for the Territory's surveyor-general. This time the experience was shared with his wife,

Joan, who joined him a few months after their first child, Penny, was born. A son, Dennis, followed in 1941 and then Amber's father-in-law, Roy, in 1944.

John was still working for the government, surveying huge tracts of Central Australia, when the Second World War broke out. After Japan entered the war in late 1941, he was tasked with retrieving official documents from the lands department office in Darwin, so they were out of harm's way in case of invasion. He made it back to Alice with the last load of papers on the day of the first air raid in February 1942, when Japanese bombers attacked the town and port of Darwin, killing hundreds of people. As the war progressed, he was assigned to survey aerodrome sites for the military, then he was appointed surveyor-general for the entire Territory.

The war was still raging when John met Ralph McKay, a Melbourne industrialist whose uncle famously established the Sunshine Harvester Works, where Ralph was engineer-in-chief before going his own way. The wealthy businessman was very interested in the Territory so John invited him to visit. When Ralph came back in 1945, they went on an inspection trip through to Tobermorey near the Queensland border and on to Mount Isa, then back again to Alice via the Barkly Tableland. Along the way, they shared camp with Bill Riley from Elkedra station, who was droving a mob of cattle to Alice.

McKay was so impressed with what he saw, and with John, that he suggested they go into partnership and purchase a station, which John could then manage. Over the coming weeks, John gave the proposal deep consideration. He'd been trying to find work with a higher salary that would fund a decent education for his children, but it was proving difficult.

By then the war was over, and his prospects were even poorer, with preference given to returned servicemen despite his years of experience.

John was reluctant to give up his profession, but he realised Ralph's offer was too good to refuse. In December 1945, he found what looked like a viable option—Lilla Creek station, south of Alice Springs. To make it work, the partners would need a regular supply of cattle to fatten, so John headed north to the popular Barrow Creek races, where he knew he would find Bill Riley.

Bill was apologetic. He'd been sick with pneumonia and had decided to sell up. As soon as the race meeting was over, he was going in to Alice to put his station on the market. John wasted no time. He wired Ralph from Barrow Creek and then took an option to purchase Elkedra, to be confirmed on inspection. In the end, Ralph and John acquired both Lilla Creek and Elkedra, with Ralph's brother, Oscar, coming on board as a third partner.

John threw himself wholeheartedly into being a pastoralist. For three years from 1946 he even served as president of the Centralian Pastoralists' Association. Then in 1948, the McKays decided to pull out of the partnership so they could concentrate on expanding their engineering business in Melbourne. Lilla Creek was let go because it was costing more to run than it made, and John bought out their shares in Elkedra.

According to Roy, his father was an incurable romantic when it came to the station that was now his own. To the west, it reached into the Davenport Range, with rocky hills and spinifex scrub. Running right through the middle was the Elkedra River, which flows after summer rain and then settles down into a series of waterholes capable of providing plentiful drinking water for cattle for up to three years if the rains don't come. Varying in size and depth, some of them are as long as 6 kilometres and more than 20 metres deep.

John decided to manage the station himself, which changed life dramatically for his family. Joan and the children had been living in Alice Springs, while he travelled between the two stations. Now they all moved out to Elkedra, where the homestead complex comprised two small cottages built by previous owners with an old storeroom between, a relatively new meat house, men's quarters and a corrugated-iron Sidney Williams hut purchased in an army disposal sale after the war. From a military staging camp set up near Barrow Creek, it was one of thousands of the iconic huts erected in the Territory to house military personnel, and then repurposed.

The hut became the Drivers' new home. It had just one long room, with exposed steel roof trusses and a dirt floor. At some stage a single internal wall was added, to separate the sleeping space from the kitchen and living space. A very good cook, Joan prepared and served all the meals, while two Indigenous girls, Jemima and Katie, helped look after the house and the children.

Initially, the children took correspondence lessons. Penny recalls sitting at a long table in the centre of the hut's living area, doing schoolwork while her mother stood at another table kneading bread dough. In 1950, Penny was sent to boarding school in Charters Towers. Dennis and Roy continued under the tutelage of a man hired to do the station books.

The boys' educational opportunities broadened in June 1951, when the first ever School of the Air opened in Alice Springs. Described as the world's largest classroom, the school today educates an average of around 110 children spread over 1.3 million square kilometres. The concept was the brainwave of Adelaide Miethke, a pioneering educationalist from South Australia, who helped raise significant funds to establish a base in Alice Springs for what became the Royal Flying Doctor Service (RFDS). During a visit to a remote cattle station in the 1940s, she was struck by the limitations of correspondence lessons and the lack of social contact outback children had with their peers. The experience inspired her to suggest teachers could deliver lessons over the two-way radio network operated by the service. After a successful trial, the School of the Air was officially opened in June 1951, by John's younger brother, Mick Driver, who was serving his final days as Administrator of the Northern Territory.

Three times a week, Dennis and Roy now adjourned to the station wireless room for 30-minute lessons to supplement their correspondence materials. This didn't last long. Dennis became sick with a rare and potentially painful childhood condition called Perthes disease, which disrupts blood supply to the head of the thigh bone, causing the bone to deteriorate. During the two years or so it took him to recover, Joan and her sons moved to Adelaide, returning at the end of 1953. The boys had only a year at home before being sent to Prince Alfred College in Adelaide to complete their education.

When John took over the Elkedra lease, it covered 1500 square miles (almost 3900 square kilometres). Under a separate grazing licence, he also had access to another

500 square miles (almost 1300 square kilometres), which contained the only two bores. Bill Riley had been developing this area on the understanding that he would be able to lease it once the government made it available. Instead, the grazing licence was taken from John, and plans were made by the authorities to lease the land to someone else. After protesting, he managed to acquire 150 square miles (almost 390 square kilometres) incorporating both bores. Another block was taken up in the name of his brother Frank.

When well-known Western Australian writer, John K. Ewers, visited in August 1950, Elkedra was carrying about 7500 Hereford cattle. This was much lower than capacity but John was being cautious. In his years traversing the Territory he had seen many fine properties ruined by overstocking, damaging the pastures and leading to wind and water erosion. He did not want this to happen to his own station, he told Ewers, who was an old schoolfriend from Perth.

Ewers was fascinated by Elkedra, writing about his visit for *Walkabout* magazine, and revisiting the piece for his 1953 book, *With the Sun on my Back*, which won a prize in a Commonwealth literary competition. He described the station as one of the greatest surprises he had encountered in Central Australia, having expected it to be mostly rock and spinifex, like much of the country he flew over to get there.

Instead, Ewers found himself sitting on a green lawn sipping a glass of ice-cold beer, looking out over a magnificent stretch of water around 1.5 kilometres long and about 35 metres wide, lined with river gums and bean trees. He learnt it was just one of twenty waterholes in a river system that started in the Davenport Range and wound its way across the station to flood plains on neighbouring Annitowa station. Over the coming days, John drove him to some of them, as well as the yards at Supplejack Bore where Indigenous stockmen were branding cattle.

Ewers's descriptions of these scenes and life in the Territory so impressed a young teenage reader living in regional New South Wales that he wrote to the author, who put him in touch with the station. Within weeks, Des Nelson was on his way to Elkedra. In his own memoir, he describes a busy life, learning to ride a horse, doing odd jobs, helping the Driver boys with their School of the Air lessons, making bore runs and working alongside John's brother Frank, who was a clever bush mechanic.

Des also spent a lot of time with the Indigenous stockmen, members of the Alyawarr nation, who lived in a camp of about 70 people some distance from the homestead. He noted carefully that they were all paid proper wages and treated well. The head stockman was Jack Spratt, who had spent most of his life on Elkedra and called the boss by his first name. He owned a portable wind-up gramophone which accompanied evening singalongs in the camp. Country music albums featuring bush tunes and yodelling were particular favourites.

Des found John to be a 'very forthright' person with an agile mind, who was a keen conservationist and strictly forbade shooting birds near the homestead waterhole. Because he enjoyed talking and vigorous debate, the boss was known as the Sandover Galah. Another popular nickname remembered by family members is the Flaming Stump—John was short with red hair. The Alyawarr people gave him the title Armugga, which means 'the arm', because he had a habit of holding one arm, with the forearm at right angles to the upper arm.

Joan, by comparison, was tall and quiet. Des's observation was that she and John were a devoted couple, with Joan providing the perfect foil to her more volatile husband. Admired by her family as a strong woman who lived an adventurous life, Joan grew up in the industrial English city of Birmingham. She arrived in Western Australia in 1929, at the age of 25, with very little, just weeks before the stock market crash that triggered the Great Depression. She was accompanied by two sisters, Mary and Margaret, with three brothers also emigrating to Australia at different times.

At least some of the family ended up at Scaddan, about 50 kilometres north-west of Esperance, where Joan made a name for herself heading up a new amateur dramatic society. Part of mallee country that John surveyed early in his career, the area was being opened up to agriculture and settled by new families, but these were tough years and many struggled to make the farms viable. With the Depression and low grain prices biting, one of the brothers rode a bike across the Nullarbor to Melbourne looking for work, while the sisters apparently moved to Perth where Joan taught typing and shorthand. According to family lore, John and Joan met on a beach in Perth, where they married in 1933.

In 1960, John died at the relatively young age of 58, after suffering for years from incurable progressive muscular atrophy, which gradually caused paralysis and confined him to a wheelchair. Only nineteen years of age when his father passed away, Dennis took on running Elkedra, where he was joined by Roy in 1963 after he finished a diploma in agricultural science at Roseworthy College in South Australia. Joan stayed on at the station until Dennis married, and then moved permanently into Alice.

The brothers worked together on Elkedra for six years, but Dennis's wife, Jan, didn't enjoy station life so he purchased a property in the south-east of South Australia. He and Jan moved there in 1969 with their daughter, Lyndie, later shifting to Adelaide and then northern Queensland. Roy was left to manage Elkedra, eventually buying out his brother's share.

In 1975, the Barrow Creek races once again intervened in the fortunes of the Driver family. That's where Roy met the love of his life, Catherine Clark. From Mannanarie in South Australia's Mid North, Catherine grew up on a farm and attended the district's little one-teacher school, where she was the only student in her year level. After going to high school in nearby Jamestown, she was sent to Adelaide to finish her education. Hating the city, she describes it as the worst year of her life.

With her mother very sick, Catherine returned home and secured an office traineeship at the local council. On weekdays she rode into Jamestown on a motorbike that her father gave her. After hours, she helped on the farm, taking shifts driving the tractor during sowing and harvest seasons to give him a break. As a reward for all her help, he bought her a car when she turned 21, and told her, 'You should go and do something else. You've done a great job.'

Willing to go anywhere but Adelaide, Catherine started checking newspapers for suitable work. One day she circled an advertisement for a position with Irish Young and Outhwaite, a chartered accountancy firm in Alice Springs. She had visited the town with two girlfriends on a bus tour just after finishing school and quite liked it, but never considered living there.

Catherine had only been working in Alice for a few weeks when she was invited to join a mob of bank employees going to the Barrow Creek races. Run over two days and incorporating a campdraft, the event was held on a dedicated track, with a striking backdrop of red, flat-topped mesas. The horses were mostly bred on surrounding properties and ridden by station hands. Spectators travelled long distances to attend, camping near the track. In the evenings many of them headed to the local pub, where Catherine met Roy on the first night.

They were six months into their courtship when Roy suggested she ask her boss, Lyndsay Stewart, if she could work through weekends for a while so she accumulated enough time off to visit Elkedra. 'I can't do that,' she told him, appalled at the idea.

Refusing to take no for an answer, Roy went to see her boss without telling her. 'I get to work this particular morning and Lyndsay opened his door and said, "I want to see you in my office!" I was thinking, "What have I done? I must have stuffed something up." Then he said, "Roy, came to see me last night." I was absolutely mortified.'

Roy and Catherine married in 1977. They were planning a quiet registry wedding but Catherine's father intervened. He wanted them to celebrate the event properly in Jamestown with family and friends, so the bridal couple made the trip, and then came straight back.

Before leaving Alice for the station, Roy thought it might be a good idea for his new wife to meet the bank manager responsible for handling the station's finances and approving loans. Cattle prices had been terrible for two years, and they weren't in good shape. 'We walked in and the bank manager said to Roy, "You might as well throw your keys on my desk cos you'll be walking off." Everyone was in the same boat; we were all broke. We had three miserable years—'77 and '78 we didn't sell a beast from here. Anyhow, things turned around in '79. Cattle prices went through the roof . . . It's either boom or bust in this country, but we survived.'

Catherine visited Elkedra a few times before she married Roy, so she had a fair idea of what station life would be like. She also spent time with Joan, listening to her stories. By then, Joan was living on her own in Alice Springs, where Catherine visited her every Friday evening for a drink after work. 'She was a good woman, a strong lady,' says Catherine. 'She was very nice but straight as a gun barrel and very direct, and I appreciated that because I like directness. We got on famously.'

At that time the Sandover was a two-wheel track, with gates to open along the way, instead of today's cattle grids. A qualified pilot, Roy had his own plane so he could fly them into Alice occasionally. Meanwhile, Catherine took on the task of driving to Alice a few times a year for supplies, taking an International truck with a large tank on the tray so she could bring back fuel, as well as anything else that was needed.

Out on the station, Catherine worked alongside Roy in the stockyards and mustering cattle. Once she'd learnt the ropes, she became solely responsible for the station's bores, criss-crossing the vast property to check the watering points and maintain equipment so cattle always had access to water.

If there was a problem with a tank, trough or fuel-powered pump, she learnt to fix it. Making the bore runs usually took about four days a week. In between she spent two days at home cooking and catching up on chores. 'I still had to cook for the stockmen, but it didn't worry me. I just did it,' she says.

Meals were served in the main homestead. Although a concrete floor and extra rooms had been added over the years, there was still no ceiling. In fact, it was possible to stand outside and throw a tennis ball between the top of the external walls and the roof into the house. To help keep insects at bay, Catherine stretched gauze over the gap. Ceilings were eventually installed but, as tends to be the case on most properties, the money-earning part of the station always came first when there was money available for improvements. At least she had a gas stove, a coolroom to store perishable food, with a diesel motor to run the compressor, and a couple of kerosene-powered refrigerators.

An old American generator picked up after the war provided 110-volt power, running for a couple of hours each night and in the morning too if Catherine had tasks like washing to do. A back-up battery supplied the lights when the generator was turned off. Before their children came along, Roy and Catherine both liked to read in the evening and play Bridgette, a two-handed version of bridge. If she was inside, Catherine might also listen to the daily 'galah session' held over the radio so station people could chat for an hour or so, sharing gossip and news, such as the latest rainfall figures. A weekly highlight was the arrival of the mail plane, which also brought fresh bread.

In 1980, the Drivers welcomed their first child, John James. Following a family tradition, his first name honoured his grandfather. His second name was in tribute to one of Catherine's brothers, who died as a baby. Then, two days after the birth, Joan died. 'She was in hospital very, very sick, and she told me she was waiting,' Catherine says. Two more sons followed after John—Clark about two years later and David in 1985. Catherine went into Alice a couple of weeks before each of them was due, keen to avoid the experience of a neighbour, who twice gave birth on the side of the road.

Having three children didn't stop Catherine from doing the bore runs—she just took the children with her. 'How they never drowned I don't know because they were free range,' she jokes. In the summer, they slept out on the front lawn, where it was much cooler than inside the house. Every evening the boys set up camp beds with fold-out legs, steel frames and wire bases, known as shearers' beds. Catherine would string a rope between two trees to support mosquito nets and the beds were placed beneath.

There was no governess, so when the boys were old enough for School of the Air, Catherine did her best to teach them. She didn't particularly enjoy the experience and it's fair to say the boys didn't either. It was a struggle to keep them in the schoolroom and focused on their lessons, when they would much rather have been outside, and the quality of the radio service for the on-air sessions didn't help. 'It was shithouse,' says John bluntly. 'You couldn't hear the teacher and they couldn't hear you, there was so much static.'

About five or six First Nations families were still living on the station then. Among them was Michael Spratt, who was born there and lived in the original homestead. His children were the same age as John, Clark and David, and they all played together. Otherwise, the Driver boys rarely saw other children apart from three trips a year to Alice for camps, swim meets and a conference for governesses run by School of the Air.

Life became more social when the station got its first proper telephone. Every Friday afternoon John and his brothers would ring neighbouring Annitowa, where a family then lived, and talk to the owner's children. Sometimes in the summer they would even come over for a swim, which involved about a two-hour drive each way over station tracks. Every Boxing Day people from neighbouring Ammaroo and Murray Downs would come to Elkedra too. Staying one or two nights, the guests would make the most of a rare opportunity to swim and go fishing for spangled perch in one of the waterholes, followed by a game of bush cricket.

However, family holidays, with all five Drivers going away together, were virtually impossible because someone always had to stay behind and look after the station. Catherine would go away with the boys, and then after she came back Roy would go away. Usually, he took what was effectively a working holiday on Beltana, an historic property between Lake Torrens and the Flinders Ranges in South Australia, which they bought in the mid-1980s. Running both sheep and cattle, it had a full-time manager. Roy liked to spend time there when he could, but it never really took Catherine's fancy.

In the early 1990s, the Drivers effectively doubled the size of Elkedra when they purchased Annitowa station, on their eastern boundary. Opening out onto the Barkly Tableland, it covers the floodout of the Elkedra River, providing valuable grazing. It's so flat in comparison to Elkedra that a tiny knoll at one bore is known as Mount Everest. Combined, the two blocks cover a million hectares, or 10,000 square kilometres, which are run as one operation. A few years later, Roy and Catherine started diversifying the herd too, introducing breeds such as Droughtmaster, Brangus and Santa Gertrudis, with the idea that it would open up more markets.

Like their father before them, when John, Clark and David reached high-school age, they were sent to boarding school in Adelaide, but this generation attended St Peter's College. John hated it. Roy only realised how much, when he caught a glimpse of his son hiding behind a tree and sobbing as he drove out of the college after paying a visit. 'We've got to get him out of there,' Roy told Catherine when he got home. They explored a few different options before settling on Emerald Agricultural College in Queensland, where John could combine a secondary education with rural studies.

Roy and Catherine had hoped their sons would spend a few years doing something else before coming back to work on the station, but just as John was finishing school in 1998, Roy was diagnosed with cancer. It had progressed so far doctors in Adelaide thought he had little time to live, telling Roy he wouldn't see Elkedra again. Determined to give him every chance, Catherine and Roy spent most of the following year in Adelaide so he could receive the best possible treatment. At the age of eighteen, John was left to run the station pretty much on his own. With his parents away and his brothers still at boarding school, it was a massive responsibility for someone so young.

Defying the odds, Roy did return to Elkedra, although he wasn't fit enough to work for a while and continued to rely heavily on John and Catherine. When Clark finished school in 1999, Catherine asked him to come home and lend a hand. Clark stayed ten years, although his heart wasn't really in it.

Surviving several major surgeries to remove tumours, Roy hung on tenaciously, then at the end of 2002 he 'got a new spring in his step'. The following year he and Catherine purchased Stamford Downs station, between Hughenden and Winton in north-western Queensland, so they had somewhere to fatten young stock that was closer to market and a wider choice of sale options. David was working in Canada for twelve months when the opportunity came up. They rang and asked if he would be interested in managing it when he got back, and David agreed.

Roy lived for another thirteen precious years. He died in January 2015, at peace because rain was drumming on the roof of the unit in Alice Springs where a palliative team from the local hospital helped the family care for him in his final days. The sound made him very happy, especially when John reported it was raining at home too.

'Roy's life was Elkedra. He often mentioned that we lived in paradise and the rest of the world was still looking for it,' Catherine wrote for his eulogy. 'Roy never wasted words. He was always considered and decisive. Challenges never fazed him. I have never seen him a beaten man. His health battle was never spoken of by Roy and never interfered with work. His steely determination and toughness got him through . . . My gift to Roy and Roy's gift to me was to love and care for each other for nearly 40 years. We had a good life together. Rest in peace my friend.'

A powerful presence still on Elkedra, Roy is often referred to in the present tense by Amber, as if she isn't quite ready to let go. 'He's still part of the family and the station, and it's probably a bit of a coping mechanism as well,' she concedes.

Amber came to live on Elkedra in early 2001. Catherine and Roy agreed she could join the station team, providing she was prepared to knuckle down and work hard. Amber and John were only twenty so there were no long-term plans. 'We were only kids. Nothing was super serious. I was just so happy to be working on Elkedra. Life was simple—it was a good time,' Amber says.

Her arrival coincided with a wet summer, making driving impossible. Instead, they relied on Roy's plane, reaching the airstrip across the river in a small flat-bottomed tinny. 'I think we had around 30 inches (760 millimetres) of rain over summer for the next two years, so for the first six months of each of those years all we did was fly the plane, it was just so wet. I hadn't experienced anything like it.'

While Elkedra is well placed to provide cattle with drinking water, even in tough years, that amount of rain is cause for great celebration. It fills the rivers and creeks, topping up all the waterholes, and then spills out over Annitowa. If it keeps raining, there is grass aplenty for the cattle for months to come. Even when it inundates the homestead, no-one grumbles, although a flood in 1977 brought more than a metre of water through the house, destroying precious photos and family documents.

The closest Amber has come to seeing anything like that was October 2005 when water reached the lawn. 'Catherine was away at the time and she was annoyed because she was missing it—it's devastating if you miss the river run. I was wondering when we were going to start lifting up the furniture and getting stuff out of the office, but the men were all so happy they were sitting on the lawn with a beer in their hands, not worried about their feet being in water, or what's happening in the house.'

But given all that water, and reasonably straight stretches of river, something else puzzled Amber. The men liked to fish and people swam in the summer. Joan Driver even had steps installed down the riverbank so she could do laps every day. But other than that, there were no water sports. Amber had grown up in a family with a ski boat and couldn't understand it. *Oh my gosh, we live in heaven. We have water without crocodiles. Where's the boat? Why aren't we skiing?* she wondered.

It took some gentle hinting, but one day Roy returned from a trip to Adelaide towing a jet ski as a family Christmas present. 'I'll never ask for anything again!' Amber promised. The family has a different model now, but summer afternoons are still often spent at a waterhole a short drive from the homestead, where people take turns on the water, or chill out with a drink before dinner. Amber's dog, Rosey, comes too.

The jet ski isn't the only gift that has brought Amber years of joy and happy memories. When she turned 21, the Killens and Drivers combined resources to fund the cost of her obtaining a pilot's licence. Building on lessons she had taken before leaving home, Amber trained for a private pilot licence in Narrabri where there is an aero club, then she completed her qualifications for an unrestricted pilot's licence in Alice Springs.

Although there was plenty of hard work and many long days, Amber's first years on Elkedra were relatively carefree. She and John shared a small cottage set on a rise about 300 metres from the homestead. There was no verandah and it needed renovating, but they were happy there because it gave them some privacy. Then in 2004, Amber and John found themselves pregnant. 'I suppose I was having a little bit of a stress because we weren't even engaged, we were just living our best life, working and doing our thing, but John's parents were over the moon. They actually said, "We can't believe it's taken this long." All we had was support.'

When it came time for the first ultrasound, John couldn't be there because he was needed on the station. However, Roy was in town so Amber insisted that he come along instead. She was very conscious he had missed out on this moment with all of his own children for the same reason John wasn't at her side. And after all, this baby would be Roy's first grandchild. 'It is one thing finding out you are pregnant, but when you actually see for the very first time another heart beat in your body, things get real! That was one of the most special moments I could share with Roy, and no amount of time will fade that memory.'

Sonny Roy Driver was born in April 2005. He was about eighteen months old when John proposed, after taking the old-fashioned step of asking Jerry's permission first. The young couple were on a bore run to Annitowa. They were pulled up at the old homestead, which is an unstaffed outstation these days. It was a far from romantic moment, and Amber was taken aback. 'I was like, "Oh my gosh, why are you proposing here?" And he was like, "Don't make me keep asking you!" And then I made him get down on one knee and ask me again.'

A few months later, John and Amber married in front of about three hundred family and friends. Amber had thought about holding the ceremony at Elkedra or back home in Narrabri, but the logistics were just too difficult. Instead, they chose an unusual venue that reflected their mutual love of Central Australia—the outdoor amphitheatre at the Alice Springs Desert Park, where the West MacDonnell Ranges provide a spectacular backdrop. Everyone then adjourned to the DoubleTree by Hilton Hotel ballroom to dance the night away.

Despite it being a meltingly hot February day, the bride looked cool and lovely in a stunning white halter-neck dress with a high waist and knife-pleated skirt, her dark-blonde hair gathered up, her smile radiant. Amber had bought the dress only a few weeks before the ceremony, during a trip to New South Wales. With no plan to wear anything special, she popped into a regular dress shop. Two assistants stepped forward to help her, under the impression she would be a guest. When Amber eventually revealed she was the bride they were horrified, immediately taking charge. Amber didn't like the gown when they showed it to her, but they kept insisting so she reluctantly took it into the change rooms. The dress was perfect.

Before the wedding, Roy and Catherine decided it was time for the young family to move into the main homestead, while they took a step back and settled in the cottage. It's a time-honoured tradition on many stations for the older generation to hand over to the next in such a way, but Amber found it both strange and daunting. 'I was incredibly nervous about the transition because Roy had lived in the house all his life . . . but they just believed it was time, and that's what was done. You couldn't say no. And literally, they packed their clothes and took their bed and off they went. It was a weird thing to process . . . but Roy and Catherine take everything in their stride. And that's one thing that they really have instilled in their boys, and I've learnt too. You just do it and move on and adapt. There's no dwelling on anything. It's always one foot after another and you keep moving forwards. That's what progress looks like. They just go with the flow.'

Her in-laws were always very open with Amber about the business and every aspect of running the station, so she could steadily learn the ropes. 'It wasn't just me being out with the men fencing and doing cattle work. I'd help in the office. I'd do the cooking. I got a good grounding in what the whole business was about. His parents are such great leaders. They're very practical how they approach things, so we always just knew. Succession was a conversation that was kind of the same as talking about the weather . . . it wasn't awkward.'

Early morning on Elkedra and the soothingly rhythmic sound of sprinklers mixes with the soft buzz of insects and the twitter of small birds in the garden. Amber is watering the swathe of lawn that surrounds the homestead. In recent years it has engulfed the old driveway that led right to the back door, and crept past a small building housing the station's kitchen-dining facilities. It's passed the old stone building that was once the men's quarters and is now guest accommodation, and is slowly spreading in front of the new quarters and along the riverbank.

It might seem a luxury in such an arid landscape to have so much lawn, especially given it has to be watered for hours most days to survive, but for Amber it's nothing short of a necessity. And not just because it cools the house by several degrees during summer, when temperatures climb as high as 50 degrees Celsius.

For Amber, the lawn plays an important role in nurturing the wellbeing of everyone living at Elkedra. It offers reprieve at the end of a long, hot and dusty day working in an arid landscape. A place where people can kick their boots off and revel in the feeling of soft, cool grass beneath their feet. Somewhere they can sit overlooking the waterhole, have a chat and reset both physically and mentally. 'It's just really nice for everyone to come back to, especially in a drought year. Even for the young fellas, who might not be able to identify what it does or understand the feeling they get,' Amber explains. 'When staff have to work from daylight to dark, it's not always roses, so if you can make a space where a young person feels they can sit down and have a yarn, how good is that?'

If Amber had any doubts about the lawn's benefits, they dissolved during the most severe drought she has experienced in her time on Elkedra. It was so dry in 2018, that the waterhole in front of the house dried up completely. Water in the nearest bore over the other side of the river was too salty, so she drove to Ammaroo station every second day to collect enough for the house, but she lost most of the lawn.

'It was really hard. When I was in town and people asked me, "How are you going?" I couldn't even talk. My throat would close up, and I couldn't say a word . . . We've had no water a couple of times, but that was prolonged. It was for a year, and everything snowballed. I wasn't able to maintain a sanctuary for people, and I was trying to deal with that and get myself through it. Not all people on stations have a lawn or a garden—it's not their thing and I get that, but for me it is. In all honesty, it wouldn't bother me if the homestead wasn't there and I lived in a tent, as long as I have a lawn.'

The silver lining was that it led to a new bore with reasonable water quality being drilled behind Catherine's cottage. Since then, the acreage of lawn has doubled in size and Amber isn't done yet, even though it already takes a full day to mow it all—a task she refuses to share with anyone because she enjoys mowing so much. And she won't hear of pop-up sprinklers either. 'Half the fun of having a lawn is you get to water it,' she grins.

Amber began expanding her green realm after Sonny was old enough to start School of the Air. She missed heading out in the morning with the men and so did Sonny. Never one to stay inside if it could be avoided, she found the homestead confining. 'My happy space is not the house—it's the outside—so I put my energies into the garden instead of doing a bore run.'

This change in the pattern of their days coincided with one of the toughest periods in Amber's and John's lives. After giving birth to Sonny, they had four miscarriages. The first happened not long before their wedding, then came two more. Determined that only immediate family and a few close friends should know, Amber did her best to pretend everything was fine. She greeted friends' new babies with a smile so as not to dim their joy, and quietly deflected questions about when she and John were planning to give Sonny a sibling. 'I just didn't want to talk about it. I didn't want anyone to feel sorry for me, or tell me to take it easy, or that it would be alright,' she says. 'I didn't want to be looked after. I just wanted to be like everyone else.'

On one occasion, Amber left a note on John's pillow letting him know the RFDS was coming to pick her up and fly her to the Alice Springs Hospital. He was out on the station. She didn't want to use the two-way radio and run the risk of other people hearing. Another time, she found herself lying to a friend. She and John had turned down an invitation to attend an engagement party because they were working, which wasn't unusual. Then Amber found herself in hospital again. She had just been discharged, when someone spotted her in Alice Springs the day after the celebration. 'I literally looked one of my friends in the face and I lied to them. I just couldn't explain.'

Amber's coping mechanism was to get up every day and work harder than the day before. Bore runs, fencing, building yards, welding, mustering, driving loaders and graders, cattle work, butchering a killer—she did them all while caring for Sonny and then came home after a full day's work and cooked for everyone, cleaned the house, tended the garden, and fed the poddies and dogs. The station bookwork was becoming her responsibility too, so she had wages and bills to pay. Then she piled on a host of volunteer roles in the community, sitting on multiple committees and helping to organise events. 'I could not say no to people if they needed a hand, I would find time to make it work. I loved the hard work and I would not have changed it for anything, but it all came at a price.'

Amber was pushing her body to its limit. She wasn't skipping meals but she lost weight; her clothing was down to size 8 and still loose. Then she made time to have an overdue pap smear. The result was 'nothing short of a nightmare'. It found cancer cells in her cervix. And so began a long series of more tests, surgery and laser treatment, with the very real risk that at the end of it all she would not be able to conceive again. Finally, years later, Amber received the all clear. With fresh strength and a sense of hope, she and John decided to try once more for another baby.

The pregnancy was in its second semester and they were thinking cautiously about telling people, when Amber flew to Brisbane to act as bridesmaid at a friend's wedding. She and Liz had gone to boarding school together and all the other bridesmaids were schoolfriends too, people she hadn't seen for years, so it would also be something of a reunion.

The day before the wedding was hectic, with last-minute errands and preparations. In the evening, the bridal party gathered at the beautiful Indooroopilly Golf Club to rehearse and have dinner. Just before she was due to practise her part walking down the aisle, Amber ducked to the toilet. That's when she realised her white suit pants were soaked with blood. The sight paralysed her with fear and sadness—a stark contrast to the happiness and excitement she'd revelled in moments before. Amber recalls a sense of time suspended. She didn't want to ruin the day for Liz, so she rang her brother, William, who lived two hours away on the other side of Brisbane and asked him to come and collect her. Then she snuck out to the car park and waited.

The bride's sister found Amber first, soon followed by the rest of the bridal party. She was absolutely mortified at the fuss when the weekend should have been a joyous celebration, all about the bride and groom. Her friends took no notice. They found a change of clothes, then the bride and her mother drove Amber slowly through rush-hour traffic to the Mater hospital. 'All I could do was stare out the window, void of emotion. I think that my heart was really broken this time. It was my fourth miscarriage and I was a world away from my family, and it could not have happened at a worse time.'

An ultrasound at the hospital showed the baby was still alive and the placenta appeared to be okay, so the doctors agreed to discharge Amber. She managed to get through the wedding, doing her best to keep smiling, but as soon as her bridesmaid duties were over she called a taxi and headed for the motel where her parents were staying. They had driven up from Narrabri, seizing a rare opportunity to see her.

The next day Amber flew back to Alice Springs. Lexie wanted to go too, but Amber persuaded her to stay behind. Not one to show emotion in public, she sat on the plane and cried all the way home. Despite the ultrasound results and her best attempts to remain positive, Amber feared the worst. Needing to be home with John and Sonny, she went straight back to Elkedra. By then the bleeding had all but stopped, and over the next few days she began to feel more hopeful.

A week later Amber returned to Alice Springs for a scheduled ultrasound, with Catherine at her side. The technician spent about five minutes taking images, then left the room. She came back with another technician, who once more positioned the handheld scanner over her lower stomach. The image Amber saw fleetingly projected on the screen was perfectly still.

Amber was immediately sent to the emergency department, where she waited for hours in a queue. In a process she had been through all too many times before, she was prodded and questioned, then told she had lost the baby. She required a dilation and curettage procedure to clear the lining of her uterus. Amber was wheeled into an empty room in the maternity ward until a surgical slot became available. A thoughtful young nurse came in to check her charts, then quietly removed the cot standing in the corner.

While the loss of all four 'tiny souls' affected Amber deeply, this miscarriage proved the hardest to bear. All the losses had been hard on John too, and placed considerable stress on their relationship. 'A man, a husband, a partner is meant to protect you, keep you safe from danger and help you when you are hurt. I was so determined to provide John with a family I didn't see how much it hurt him not being able to make me better, or keep me from the surgeon's theatre, or even be with me when I was in hospital. There was so much I had to do without him, and John without me,' she says.

At this low point in their lives, John and Amber decided it was time to focus on the positive and be thankful. They had a beautiful boy and she had survived cancer. Realising that it wasn't fair on anyone to keep trying, they decided the next attempt to have a child would be their last.

Ruben John Driver was born in July 2011, and this time John was by her side. Ruben had her devoted attention from the start. 'For the first twelve months of his life I felt like I didn't move. When Sonny was born, the second I got out of hospital I was working, but with Ruben I just let myself enjoy having a baby, so if I didn't want to go out and do any work I'd just sit on the couch and look at him.'

A mild June morning and Amber is making her way to Demon's Hill bore, on the western edge of the station. It's recently been upgraded with a corrugated-iron tank that holds more water and a solar-powered camera system that takes snapshots of the troughs at a certain time every day. Amber can view them back in the office to check for signs of trouble. On some of the most remote tanks there is a telemetry system that monitors water levels in the tanks too. 'It's a good tool if there's a lot going on, or we don't have a lot of people to do bore runs,' she explains. 'This time of the year cattle can last a day or two without water, but it's not something you want happening in summer.'

While the monitoring systems help avoid emergencies, there is no replacing experienced eyes and what they can detect on a bore run. 'It's something we explain to the young fellas. Don't just drive with the blinkers on. Keep an eye out on the fence line and what the cattle are doing,' Amber says.

Only the evening before, Ruben proved that he has already absorbed this lesson at the age of ten. Driving his buggy along a track near the homestead, with his dog Scooter in the back,

Ruben spotted a heifer sitting down. Unusually, the cow didn't move as the vehicle approached so he stopped to check if she was in trouble. Following behind, Amber took quiet pride in Ruben's actions. 'You don't want to be telling them what to do all the time, especially if they want to come back,' she says.

Today it is Amber's turn to pay attention. She is driving one of the station's sturdy four-wheel-drive utility vehicles, which she fuelled up and checked over carefully before leaving. Years of mentoring by old bush hands and plenty of practical experience have reinforced the importance of being prepared in this isolated and sometimes treacherous country. The vehicle is loaded with paraphernalia that might come in useful if she needs to repair a pump, mend a fence or replace a flat tyre. Solar-powered pumps are gradually being introduced across the station, but she is also carrying fuel to top up supply for an old engine at one of the bores. She's also packed smoko—small pies made with leftover roast meat, and slices of banana bread and chocolate cake.

Passing the station workshop, Amber comes to the stockyards where most of the weaner cattle are drafted. It was the scene of one of her more unfortunate experiences on Elkedra. She was working with John and some station hands to process cattle. Among the mob was a troublesome beast that kept charging around the yard, forcing people to escape over the fence. On her last jump, Amber landed on a star picket which went straight up through her jeans, into her knee. 'A heap of the fellas came over to see why I was still sitting there, and I didn't want to show them because it was a few weeks since I'd shaved my legs, and I wasn't pulling my jeans up for anyone,' she confesses sheepishly.

John wasn't having it. Ignoring her protests, he exposed the leg, took one look at the injury and carted Amber over to the homestead where Catherine packed the wound. Then she returned to the yards and went back to work, this time assigned to keep the tally book and load the eartag applicator. 'We had a couple of thousand head of cattle we needed to deal with,' she explains matter-of-factly.

The yards sit at one end of a 30-kilometre laneway that embraces the Elkedra River as part of an extensive trapping system. Most of the time cattle walk freely through narrow gates to access water when they need a drink. When it's time to muster, the gateways are fitted with short pieces of steel pipe, forming a row of arms on both sides that fold back as the cattle push through. The arms only move in one direction so the cattle can't go back the same way.

Roy was the first station owner in this part of the world to introduce the system; it was part of a major development program he began at Elkedra in the 1970s. There was only one set of permanent stockyards when Catherine arrived. Now there are many kilometres of laneways and multiple permanent yards that make mustering and managing the cattle easier.

One of the longest laneways stretches along the George, a creek that feeds into Elkedra River not far east of the homestead. A cluster of trees on one of its banks was chosen as a camping spot by a surprising visitor in 2011. Famous Australian actor Jack Thompson worked on Elkedra in the mid-1950s, when he was only fifteen. He spent about a year on the station working as a jackaroo alongside the Alyawarr stockmen, camping in the bush and learning about their culture. It was a formative experience in developing his appreciation of Indigenous cultures and the outback, which has coloured the roles he's played over a long and illustrious career. Elkedra was so important to him that he brought his partner, Leona, back to camp on the George for a week or two, with the Drivers' permission.

Mustering still takes months on Elkedra and Annitowa, but the trapping systems save time, effort and money, and reduce stress on the cattle. The first crew of contractors usually arrives in mid to late June, and the last in about November, depending on how quickly surface water from the previous wet season dries up, forcing cattle to enter the traps for a drink. Helicopters are used to locate and bring in the stragglers. There are no horses on the station anymore, so Amber regretfully packed away her saddle, but many of the contractors bring them, or use motorbikes. The Drivers prefer permanent staff to stick to buggies or one of the Toyota utility vehicles, which are safer in rough country they don't know well.

When Amber first came to Elkedra, cattle were mustered at every watering point. She and John would camp out with the crew, even if it was only 20 kilometres from the homestead. 'Our lives were spent in the swag,' she says. That changed for Amber when the children came along, and now even John comes back to the house most nights.

In charge of cooking for the crews, Amber makes sure the freezer is full of baking for smoko, and that there is bread and cold meat for sandwiches. The resident governess usually lends a hand. They may even cook the evening meal and bake, if they enjoy it and are competent. Meat is the main component of just about every dinner, so every governess leaves Elkedra with some knowledge of butchery. A door off the kitchen leads to a large meat room, which has its own sink, an old commercial meat and bone band-saw, meat hooks suspended from iron railings, and several freezers.

The Drivers slaughter and eat their own beef and sometimes kill a pig they have raised. They corn meat too and make their own sausages, which can turn into a weekend event, shared with friends and neighbours. Everyone digs in to make about 800 kilograms of mince, which is divided into 20-kilo batches and then flavoured. Curry powder, bacon and onion, spinach and pine nut, jalapeno and chutney are among the favourites.

Most of the time, the men are around for lunch. They grab a quick breakfast together before six o'clock, taking a few minutes to make their own sandwiches and choose from a range of cakes for smoko. By the time they head out for the day, Amber is watering the lawns and tending the garden. The governess is generally given a reprieve from these early starts, as long as she is in the schoolroom with Ruben, ready to start lessons by eight o'clock.

One of three permanent employees on Elkedra for much of 2022, Gemma Watt is a Sydneysider who has just signed up for her third year on the station and is highly valued by the Drivers. It's not always easy to find a capable governess who enjoys tutoring and can cope with the isolation, and it's even harder to keep them if word spreads. 'We are so lucky!' says Amber.

Before heading out, she asks Gemma about Ruben's lesson plan for the day. It's getting towards the end of term, so they are concentrating on catching up with outstanding items on the School of the Air curriculum. They are also preparing for one of Ruben's new teachers, who is coming out to meet him in person. It will be her first visit to a station.

There are two particular exercises Gemma and Ruben are looking forward to tackling. The first is a writing exercise that asks Ruben to cook a batch of popcorn. He then has to write down words that describe what he has heard and smelt. They will be given to someone who wasn't there to see if they can guess what is being described, testing how well he has done.

Another teacher wants them to use Oreo biscuits to learn about the different phases of the moon. Ruben has to break the biscuits open, then shape each one so the cream fillings replicate each phase. The task is clearly going to involve chomping on the biscuits to get the right shapes. 'A highly creative teacher wrote that unit,' Amber reflects. 'A kid who might not be interested at all in the moon is going to be highly motivated.'

The coming afternoon's 45-minute online session will be an art lesson, something that would have been difficult before satellite dishes, the internet, computers and webcams revolutionised the way students learn. Now they can see their teacher, and each other, not just hear them. However, just as in John's day, there are technical challenges. Issues with bandwidth mean Ruben's teachers have to restrict the number of students joining via webcam or the system crashes.

While Gemma and Ruben knuckle down, Amber is driving down the two-wheel track leading to Demon's Hill, her vehicle rattling and bumping over relatively flat terrain. She spots a cow with a newborn calf. They seem to be doing fine so Amber drives on towards one of her favourite places on Elkedra, known as Old Station. There is little sign of it now, but there used to be a stone dwelling, set on the slope of a rocky ridge dappled with spinifex and small scrubby trees. Native apricot trees grow lower down, closer to the river, where Amber climbs over huge rocks for a view of the waterhole.

Further down the track, she stops at a large, scattered patch of scrubby plants carrying small delicate pink flowers with deep purple hearts. Sturt's Desert Rose, a member of the hibiscus family, is the Territory's floral emblem. They grow all over this part of the station. Amber picks a few to take back to the house and then drives on to her favourite tree, where she pulls up for smoko. A sprawling ghost gum with a glowing white trunk and welcoming branches, it's only a few metres off the track. Whenever she's passing, she likes to stop and say g'day.

A bit further along, Amber passes a rocky cliff that provided a vital clue in solving a mystery. After Roy's death, someone from the primary industries department visited the station looking for two sites photographed during drought in the 1960s to monitor flora and fauna. They had photos on file but not precise locations. One of the images showed the rocky cliff, which Amber recognised immediately. The nearby creek line had changed but they eventually found a picket marking the spot. The other site involved more guesswork but it was eventually located too.

Seeing the station regularly from the air gives Amber a clear picture of its main features and landmarks, but that doesn't mean she always knows exactly where she is on the ground. She can't quite bring herself to use the word 'lost' but there are times when Amber has been 'strategically misplaced'. She was out on a bore run one day with Emily, who came to work on the station when Sonny was about six months old and stayed for around four years. 'It was a good start to the wet season and we'd had a lot of rain. We got into the road and it was like driving in a channel for about five or six ks. I couldn't drive up the sides and I couldn't go backwards so my only choice was to keep going forwards. We spent all day digging ourselves through a bog, and then we had to turn around and drive back again. What was going to take a couple of hours took much longer and we didn't have lunch with us, so I made a strategic decision to leave the road and follow some hill lines and a valley.'

Where they ended up was not exactly where Amber expected. Fortunately, she soon recognised a landmark and they made it home okay but it was a reminder of two golden rules when it comes to travelling in the bush. 'The rule of thumb back then was that if you are not back by beer time, when you should have been, then someone will come and find you, so stay with your car and wait. And it's a good idea to be where you tell people you are going to be so they know where to look. That day we were potentially in a bit of hot water because we were not on the road we should have been on.'

Apart from mechanical breakdowns, the issue that most commonly prevents someone in the family from reaching home is water. Half an inch of rain is enough to make some roads impassable, and creeks can fill with little warning. On one occasion, John woke her up at two o'clock in the morning to say he was stuck on the main access road. Amber was not sympathetic. She had warned him earlier in the evening that it was raining at Elkedra and the creeks were up. 'It wasn't raining where he was, and he perhaps thought I might have been exaggerating or didn't know what I was talking about so he came home anyway.'

Feeling cross, she told him he would have to wait until daybreak. 'I will get you tomorrow when I'm ready to come and get you,' she said. Knowing that Roy and Catherine would have heard the conversation on the radio receiver in their cottage, she marched over to see them. 'Roy is sitting there pulling his boots on. You can't tell Roy not to do something and he knew I wasn't going to go, so off he went. He's a good dad isn't he, looking after his son!'

Another time, John came to collect Amber from Ammaroo station where she was visiting their nearest neighbour and dear friend, Anna Weir, who lives at the end of what might be considered Elkedra's driveway. The 85-kilometre trip over was fine, but by the time they returned one of the creeks was running a banker. They stayed in their vehicle overnight hoping the creek would be passable in the morning. When it wasn't, Anna came to collect them in her helicopter.

As dusk approaches, people already back at the house keep a watch on returning vehicles to make sure everyone is accounted for. It's a longstanding tradition on Elkedra for all the family and staff to come together for the evening meal, even if that means squeezing up to thirty people around the table during muster season. 'I really don't like it when people miss out, so we make it work,' Amber says.

Until a few years ago, meals were served in the homestead, then the plumbing system packed it in. Drainage from the kitchen sink failed first. For a while Amber made do, using the laundry for water and to wash up, then the drains in there blocked up too. Investigations showed they were unfixable without major renovations. The homestead is a ramshackle old building with many structural problems; there's a running joke in the family that if the termites didn't hold hands, the house would collapse. Given the expense of bringing tradesmen and materials out to Elkedra, John and Amber figured it wasn't worth the effort.

A better option was putting up a new kitchen and dining area, attached to the existing meat room a few metres across from the back door. It's a welcoming space with air-conditioning where staff can come and go as they please, without worrying about disturbing the family or taking off their boots. The L-shaped open-planned room has white corrugated-iron walls and large windows overlooking the garden. The floor is covered with large terracotta-coloured tiles that are easy to clean and hide the red dirt. A large central stainless-steel bench with open shelving serves as the main workspace in the kitchen area. Amber does not claim to be a master welder, but she made this bench and the dining tables too. They fill half the room, while a small occasional table tucked in a corner serves as the 'post office'.

A large white metal box containing emergency first-aid items sits on top of another bench, below a large map of Queensland. In the opposite corner a small television is fixed to a bracket set high on the wall, just inside the main door that leads out into a small patio area which serves as a beer garden. Everyone gathers here in the late afternoon to share a drink and catch up on the day's news before dinner is served at seven o'clock. 'If you only want to have a cup of tea that's fine,' Amber explains. 'The point is that you come together. You can talk about what went well that day, or what didn't, and what needs to be sorted out tomorrow, or tell a tall tale.'

Within a year or two of moving to Elkedra, Amber began volunteering with various organisations. In part it was about creating opportunities to meet people and immerse herself in a community she was fast coming to love. Today it is more about giving back and doing what she can to encourage the next generation of teenagers dreaming of a future in Central Australia. Sometimes it has involved finding her voice and stepping into the limelight to fight for better services, or to protect the beef industry that keeps most stations afloat.

In 2021, Amber and Sonny agreed to be part of publicity for the RFDS, to raise awareness about the importance of the service to outback families and generate more donations. They spoke from the heart, in deep gratitude for the assistance Sonny received during a medical crisis that might have cost him a leg.

He had been back from school for less than 24 hours when he crashed his motorbike. Sonny and Ruben love riding their bikes, whether it's out on the station or in motocross and enduro competitions. In fact, it's something of a family pastime, along with entering their bright green 1977 Holden Kingswood HZ wagon and 1982 silver Statesman Deville in the Red Centre NATS event in Alice Springs every September. Like most station kids, the boys learnt to drive at a very young age, and they have been riding bikes even longer.

On this particular day, Sonny and Ruben were only a few

kilometres from the house. Swathed in clouds of dust, they had no time to react when their bikes emerged from different directions and collided with terrible impact. Ruben picked himself up but Sonny didn't.

Amber was a bit concerned when she heard only one bike coming back. In his rush to get the words out, it took Ruben some time to explain what had happened. As soon as his parents saw Sonny, they knew it was a case for the RFDS. One leg was clearly badly broken and he was in tremendous pain. 'His leg was the standout problem—it wasn't where it needed to be. At that stage, he could feel his fingers and toes, so we ruled a few things out, but I didn't know what, if any, internal things were going on,' Amber explains.

Leaving John to sit in the dirt alongside their distressed son, Amber rushed back to the homestead and called the RFDS. Taking a deep breath to steady herself, she explained what had happened and described her son's injuries. An aircraft carrying a retrieval doctor and flight nurse was soon on its way from the base in Alice Springs. Meanwhile, one of the medical team gave Amber detailed instructions about the first aid she needed to give.

The most important task was straightening Sonny's damaged lower leg. That meant manipulating it without being able to give him painkillers. 'Pulling a broken leg straight and telling your kid to suck it up is not for everybody,' Amber says with some intensity. 'I feel emotional talking about it now but at the time that was something I had to turn off. My mum emotions were not there and it had to be like that so I could deal with it. It's sort of like you are a different version of yourself. It's a very strange thing to experience.'

David was waiting at the station airstrip when the plane landed, and quickly drove the medical crew to the accident site. Working quickly and calmly, they gave Sonny medication for the pain and stabilised the leg to prevent further movement. Amber flew with her son to the Alice Springs Hospital, where further investigation revealed that he was in very real danger of losing his leg. Around 5 centimetres of bone in his tibia had shattered and he was suffering compartment syndrome. In reaction to the trauma, the muscles in his leg had swollen, compressing the nerves and blood vessels, and blocking blood supply. It was now a surgical emergency to ease the pressure and avoid amputation.

More surgeries followed over the next few weeks. For most of that time, Amber didn't leave her son's side. 'She was there night and day for pretty much two weeks while I was sleeping all day and waking up in the night screaming with pain. Without Mum I don't reckon I would have done that well,' Sonny explained after his recovery.

'I cannot describe the incredible comfort that comes from knowing that you can simply pick up the phone and talk to someone at the RFDS, and know that expert caring help is always at hand,' Amber told a journalist.

Another outback institution that has been close to Amber's heart since her first child reached school age is the Isolated Children's Parents' Association (ICPA). The ICPA is dedicated to ensuring geographically isolated children are not disadvantaged because of where they live, when it comes to their education. They are a strong voice for outback families, listened to by governments on issues such as communication infrastructure, early childhood services, curriculum policies and affordable boarding options. After learning what the organisation was about and the difference it made to the quality of education children such as Sonny and Ruben receive, Amber did not hesitate to get involved. 'I remember thinking, "This is incredible." Their voice is so well respected, so many women and hours of volunteer work, and they have the ear of members of parliament in Canberra.'

Amber spent ten years on the executive committee of the Alice Springs branch, and helped organise the 2017 Federal ICPA Conference which it hosted during her first year as president. She also served on the Northern Territory State Council of the association for eighteen months, spending what amounted cumulatively to weeks away from home attending meetings and advocating with others on behalf of her community.

Amber says she has never been shy about throwing a spotlight on issues affecting Central Australia. In fact, she has become quite fierce, particularly about the poor state of roads. Vital routes such as the Sandover Highway are often so bad, they have an impact on the quality of life for outback families and have significant economic consequences for their businesses. In 2021, Amber calculated conservatively that it cost the station around $26,000 in extra transport costs to get Elkedra's cattle to market and around $70,000 in lower prices because of the amount of condition cattle lost along the way.

Like many others, the Drivers are also frustrated about poor telecommunication services. The landline at Elkedra has been so unreliable for so long, that after years of complaints and promised repairs failed to improve the situation, they built their own 4G small-cell mobile tower. It was expensive but essential for both the business and keeping young staff, who see being able to use their smart phones as essential, even out here.

Amber has even screwed up her courage to take on the mining industry. In 2017, she wrote a submission on behalf of Elkedra and sent it to the Pepper Inquiry. Led by Justice Rachel Pepper, the independent scientific inquiry explored the potential environmental, social and economic risks associated with fracking for gas in the Northern Territory.

In her submission, Amber explained the unmatched ability of stations in Central Australia to produce organic beef because of the way cattle are raised naturally. She was deeply concerned about the potential damage to the industry and its reputation for producing 'clean and green' meat if precious water sources were polluted by the chemicals used in fracking. Allowing it to go ahead would be the equivalent of gambling with her family's future livelihood, for generations to come, she told them.

The submission was brief and from a layperson's perspective, but it had considerable impact. A few months later she was asked to attend a panel hearing in Alice Springs. Thinking it was just going to be an informal conversation, Amber was shocked when she arrived at the venue to discover a room full of people and a camera recording what turned out to be formal proceedings. 'It was a little daunting,' she says. After listening to what Amber had to say, Justice Pepper thanked her for the presentation, which she said was important; Amber was the first person to raise these particular issues with the panel, and they were 'now very interested' in following up the matter. Released in March 2018, the inquiry's final report made 135 recommendations to mitigate risks involved with fracking.

The Drivers have also gone head-to-head with individual mining companies wanting to come onto Elkedra and search for minerals and petroleum. Until 2020, landholders in the Territory had no legal right to veto access and the Drivers had some bad experiences. The worst example happened about twenty years ago when an exploration crew drove onto Annitowa to access water without permission. They left gates open and wrecked a bore, which collapsed. John tracked down who was responsible and, after extensive negotiations, the company put down a new bore and equipped it. 'That's when I thought there has to be more to protect us,' Amber says.

As a result of the Pepper Inquiry, new regulations came into effect in 2020 requiring petroleum companies to reach formal agreements with landholders in the Territory to access their land. These legally binding documents set out rights and obligations that must be met when a company enters a property to search for petroleum, and compensation rights for wells drilled and any damage caused.

Amber already knew about the concept of land access agreements, and had been asking companies to sign them for some years. Her terms usually involved notifying the Drivers before coming onto the station, maintaining or repairing station roads used by their vehicles, and measures to protect bores and essential water supplies for cattle. Most companies agreed to sign with little pushback.

There was one notable exception which has Amber shaking her head to this day. A Canadian-owned company planned to drill exploratory wells over some of Elkedra's best grazing land. Its senior executives refused to agree to Elkedra's usual terms, presenting their own for the Drivers to sign. Amber refused. In an attempt to resolve the situation, the company flew senior executives all the way from Canada to Elkedra. In a surreal moment, the businessmen lined up on one side of the table in the kitchen–dining room, while John, David and Amber sat on the other. 'Where is the boss?' one of the company men asked. 'You're looking at them,' Amber replied.

The businessmen laughed and then realising the Drivers were serious, accused them of wasting their time and shareholders' money. They insisted the Drivers sign immediately, pushing the paperwork across the table. Amber politely refused and pushed it back. After arbitration failed to resolve the dispute, the matter was due to go before the Northern Territory Civil and Administrative Tribunal in what would have essentially been a test case for the new regulations, but the company apparently thought better of it and dropped their intention to carry out explorations on Elkedra. 'I think we must have been too much of a hot potato,' Amber says.

Looking back at these experiences, Amber recalls what happened when she was in Year 11 at Calrossy and contemplating running for school captain. She didn't like public speaking and the role involved plenty of it, so she decided against standing for election. 'Being afraid of something held me back from reaching my potential. Dad always said to me that I should never be afraid to speak up. It doesn't matter who they are or where they've come from, or what they're telling you, people are just people. You are just talking to another person,' Amber explains.

'This is about our lives and our livelihood. These people might quit their job tomorrow as a director of one of these companies and go and work somewhere else, but I am protecting our home and future generations, so I can be just as fierce. There is no doubt that we need minerals and gas, but it can't be at our peril. We are not anti-mining, or anti- anything, but we need to raise awareness that you can't decimate an industry. For me it's a matter of finding the courage to say what I need to say. That's a big part of it—being courageous to stand up for what you know.'

Now she is in her forties, with a son fast approaching the age that she was when she set off for the Territory, Amber is becoming increasingly passionate about doing whatever she can to help build a more resilient future for the Central Australian community. That means encouraging young people interested in agriculture and making it easier to live in isolated places like Elkedra. 'I'm just so excited about the really good people coming through and the potential they have to keep the good work going, but we are only going to attract young leaders if we have services that support families.'

Her own family's resilience is on Amber's mind too. She and John are trying to find a better work–life balance. Feeling burnt out, she has stepped back for a while from some of her volunteer commitments. Ruben will be going away to boarding school soon, so there might even be more time to spend gardening or sewing quilts, and making a few more improvements around the homestead complex. In 2021, the Drivers decided to spoil themselves and had a reverse-cycle air conditioner installed in their own house. 'There's no going back now mate!' she told John.

Air-conditioning has also been installed in the guest quarters, to entice members of her family to visit more often. Jerry has just turned 80 and Amber is more conscious than ever that every moment counts. 'And that's one of the biggest challenges, living remote—I can't just pop over to see them and say g'day,' she frets.

Lexie is more philosophical. Coming back from a visit to Elkedra a few years ago, she turned to Jerry and said, 'I haven't thought about this before, but Amber is in the perfect place . . . She doesn't need us anymore. She still wants us, but she doesn't need us, and there is a huge difference.'

Lexie's conviction that her daughter has found her true path in life was reinforced some time ago, when Amber came home to visit and spent time sorting out boxes of childhood possessions. One day Amber called her into the room. She was sitting on the floor cross-legged, reading something she had written in primary school. 'Listen to this, Mum. "When I grow up, I will own a cattle station."'

2

A Little Bit Sheep Crazy

KELLY DOWLING, DALTON, NEW SOUTH WALES

Shearing is over and the 100-year-old woolshed on Denbeigh station stands silent. Golden shafts of light pierce the gloom of this bush cathedral dedicated to the worship of wool, spotlighting its hand-hewn timbers and lanolin-soaked boards. Made from the trunks of stringybark trees, some of the support posts are so huge that even the lankiest shearer would stretch to hug them. They were dragged into place by bullock team, and hoisted into holes wide enough to hold three men.

Generations of the Dowling family have toiled in this woolshed, which stands on a rise overlooking the Lachlan

River. When its timber shutters are thrown open, the view is one of Kelly Dowling's favourites—not just the rolling hills of the Southern Tablelands or the sweep of river gums growing along the meandering river, but the interior scene too. Holding pens filled with the merino sheep she has bred and nurtured. Shearers skilfully stripping away the miracle fibre that has sustained the Dowling family for more than 150 years.

Kelly handles every single fleece, classing each one before it's pressed. A relatively straightforward task but the scale of this personal commitment is astonishing. When she came back to work on the farm around twenty years ago, it encompassed one main property and a smaller holding running 10,000 sheep. Thanks to her drive, the support of her family and a close-knit circle of friends, Denbeigh is just one of five properties now in Kelly's charge. Between them they run 36,000 sheep, and some of them are shorn twice a year.

A self-confessed 'control freak' who loves setting goals, Kelly has harnessed her natural strengths and lessons learnt in army officer training to push the Dowling enterprise beyond anyone's expectations, except, perhaps, her own. Motivated by a genuine passion for wool and a deep love for her family, she has pushed herself too, overcoming life-threatening illness and heartbreaking loss to create a legacy that fills her father with enormous pride. A man of few words, when asked about his daughter, Eric says simply: 'She can do anything.'

Born in 1975, Kelly grew up in the district of Dalton, where Dowlings have made their home for six generations. About an hour's drive from Canberra and not far from Gunning, this is traditional sheep-grazing country, and many of the original pioneering families still hold ground. Fascinated by her own ancestry, Kelly recounts the story of her three-times great-grandfather, Robert Dowling.

From English farming stock, Robert sailed into Sydney Harbour in 1831 as a crewman aboard the *Nelson*, which had been hunting whales in the South Seas. Ten years later, he married Amelia Horton, who also came from England. They must have seemed an odd couple—Robert was 37 and short with a round face, while Amelia was tall, angular and twelve years his junior. They settled in the Dalton area where nine children were born, including Mary Ann Dowling, whose son, Sir Walter Merriman, was the famous Australian merino breeder.

Mostly a farmer, Robert was also an innkeeper. He built the Dalton pub, which he named The Wonder of the World Hotel. He also established a bush racetrack and inn on his pastoral property north-west of the village. The inn's ruins still stand today, well placed on what was once the main road connecting Gunning and Burrowa (now known as Boorowa).

Before too long, Robert bought the neighbouring land where Kelly now lives and named it Coolong. The property was passed on to his eldest son, another Robert, who made his home there with his wife, Mary Ann Amelia Bush. The couple had an astonishing seventeen children, with all but three living to adulthood. Two died as infants while their eldest son, William, was killed in a tragic accident at the age of twelve when he fell off a load of wood and was crushed by the wheel of the dray. According to a story handed down through the family, his distraught mother took a spade to the spot where he died and gathered up his blood so it could be buried with him.

Described as an astute and cautious businessman, Robert junior purchased Denbeigh in a mortgagee sale in 1898 and sent two of his younger sons, Phillip and Loral, to live there. The property was almost 30 kilometres by road from the Coolong homestead, in the Bevendale district. Phillip was only sixteen at the time, and Loral, who is Kelly's great-grandfather, was fourteen. Their older sister, Rachel, occasionally stayed with them, taking a shortcut over the hills so she could act as their housekeeper, but on the whole, they apparently managed alone despite being only teenagers.

The boys lived in a solid stone house built by the previous owners, with cedar-framed windows facing the Lachlan River. It's derelict now, but Kelly loves to stand in what used to be the kitchen, looking out through the missing windowpanes and trying to put herself in the shoes of the settlers who made it their home. 'They would have valued that view. It's so precious, nature at its best, and I don't know, maybe their lives were easier in many ways,' she says.

In 1923 Loral built a stylish new house within metres of the old home to better accommodate his wife, Winifred, and their eventual five children, including Kelly's grandfather, Horace, who was the oldest. A standalone dining room was also constructed across from the back door, making it easier for Winifred to cater for shearing teams and station hands. In the same year, Loral built the woolshed that Kelly loves so much, using his own bullock team and draught horses to drag its giant support posts up to the site.

Even though he died more than 50 years ago, Loral looms large in the Dowling family memory. Nicknamed Crikey, Kelly's father, Eric, remembers him as a big solid man, although not as big as Robert junior who weighed around 100 kilograms at the age of sixteen, and 140 kilograms by the time he was twenty. 'Crikey was a pretty amazing character,' Kelly reflects. 'He died before I was born but the stories people tell about him . . .'

One of her favourites relates to Dalton's famous fossil rock. Dating back to around 21 million years ago, it carries finely detailed impressions of leaves. The rock was used as a door-step in a local store before the council took charge of it and placed it in a town park. 'Crikey put it on a truck and took it away because they didn't put a shelter over it to protect it, so he just stole it and told them he would only bring it back if they looked after it.' The shelter was built and the rock duly returned. 'He was a bit of a scoundrel I think,' Kelly says, with a softly snorting laugh.

Each of Crikey's three sons was eventually given their own farm to run. Eric's father, Horace, known as Hobby, was allocated a property just over a hill from Denbeigh. This is where Eric spent his youth with his older brother, Colin, before heading to Canberra. As the youngest son, he intended to build a future off the farm, and found work at the Bureau of Agricultural Economics.

Not long after, he met Kim, who was just one week into her training as a nurse at the Royal Canberra Hospital with Eric's cousin Glenda. Eric turned up at a nurses' dance where

Kim spent most of the night partnering one of his best friends, but they got talking after the dance ended and something clicked. He was soon coming to see her every weekend when she wasn't rostered to work, and sometimes during the week if she had a day off. 'We'd stay up all night and go to work at seven the next morning. When I think what I put my patients through!' Kim laughs.

Kim and Eric married in September 1973. Three months later, tragedy forced their lives to take a very different path when Colin was killed in a vehicle accident. They swapped their Canberra apartment for a small house built alongside Eric's parents' home, and Eric became a full-time farmer. Looking back, Kim believes it was inevitable they would return to the property one day. 'I just think it was in Eric's heart,' she says.

A farm girl born and bred, who loves horses, Kim was more than happy with their change in plans. The eldest of four children, she grew up in the Grenfell district where her father's family were highly respected sheep breeders. Much like the Dowlings, the Bembricks had been farming the same patch of ground since the 1860s. In 1892, they founded the Wirega Merino Stud which was passed down through the generations to Kim's grandfather Bruce and then her father, Brian. Evidence of their success hangs in a small building at the bottom of Eric and Kim's driveway, where Brian established a private museum in 2005. It houses a treasure trove of records and artefacts accumulated over many years, including numerous trophies and a blanket stitched together out of prize ribbons. Among them are more than a few champion sashes from the Sydney Annual Sheep Show.

Collecting historic items was a passion for Brian. He filled a shed at Wirega before retiring with his wife, Diddie, into Grenfell, where he built and filled another shed behind their house. When Brian and Diddie decided to move to Canberra, Eric and Kim proposed housing the contents in a small building that had served as part of their shearers' quarters. Brian spent weeks moving and unpacking the collection, displaying it on cupboards and tables bought at local clearing sales. As word spread, people began to contribute more material and the museum soon sprawled into every shearer's hut.

Brian died in 2020 a few weeks short of his 89th birthday but the family still keep a watchful eye on it, opening the collection up by appointment. Visitors are fascinated by the eclectic mix of items associated with rural life, from one of Loral's bullock yokes, a homemade shepherd's crook, shearing handpieces, rabbit traps and farm tools; to kitchen equipment, pantry items and Brian's childhood teddy bear. A small set of shelves contains a sample of handmade bricks manufactured on Dowling land in the colonial era, including a series each indented with the symbol of a playing card suit. The earliest bricks used on Coolong were convict-made and stamped with a heart.

Handwritten labels prepared by Brian identify every item, often revealing his quirky sense of humour, along with his atrocious spelling. 'Pop Brian was the worst speller and he was very funny,' says Kelly, pointing out a collection of tobacco tins that he labelled 'Cancer Corner'.

Looking back, she appreciates the number of years Brian was part of her life, and her father's parents too. 'I was very lucky, in a sense, because I always had grandparents around me. I'd be sent off on holidays for a week with Nan Diddie in Grenfell or Nanna at Bevendale,' she says. 'Hobby was gorgeous. I used to get on with him famously, and with Nanna. They were both a big part of my life, and I was Pop Brian's favourite because I was the farmer girl and I liked sheep showing.'

Even when she was a baby, Kelly spent most of her time out on the farm. After seeing her training through to the end, Kim gave up nursing and helped on the property, working alongside Eric and Hobby, while her mother-in-law, Una, focused her energies on community work. Born a Shepherd, from Wheeo station near Crookwell, Una was a strong and independent woman ahead of her time, who ran everything, according to Kelly, from the Bevendale church and tennis club to a local branch of the Country Women's Association (CWA). She was also an enthusiastic gardener, following in the footsteps of her mother, who won hundreds of prizes and trophies in the flower section of the Crookwell Show, and more than 500 prizes for show cooking.

With her mother-in-law so busy and unable to babysit, Kim often took Kelly out in the truck when she was mustering sheep, or into the shed when they were shearing or marking lambs. She did the same with her next child, another Robert, known as Rob or Robbie, who was born in 1978, when Kelly was three, and then Luke, the youngest sibling, born six years later. As they grew up, all three children were put to work on the farm too. Kim recalls Eric urging them all on when they were doing sheep work, with promise of a holiday to keep them motivated. 'We lived a very rural, easy life—no holidays, no money, but the best,' Kelly says.

About the time Kelly started school, Eric and Kim bought their own property, Tarcoola, between Denbeigh and Coolong. This is where Kelly spent the rest of her childhood, riding horses with her mum and playing happily with her brothers. The three siblings were close, despite the age gap, although she was particularly close to Rob. Kelly recalls hours spent playing tennis and swimming in the Lachlan River, which runs over a stretch of sand below the house.

The highlights of every year were picnic gatherings on Anzac and Boxing days with friends and neighbours, the Hallams—an ongoing tradition. In those days, the two families usually gathered on the Hallams' property at a popular swimming hole the size of an Olympic pool, created by sand mining.

When friends came to visit Tarcoola, a favourite pastime was pedalling bikes to a nearby rocky outcrop covered with gums, where two small caves look out over Blakney Creek, a habitat for platypus and water rats which never runs completely dry. Sometimes Kelly went on her own, sitting in one of the caves to write. As a child she thought it was a huge space, but in reality it is only just big enough for a moderately tall adult to stand up in.

Comparatively small for her age, with long golden-blonde hair, Kelly spent the first six years of her schooling at Dalton's one-teacher public school with only a dozen other pupils. Every morning, she rode her little red bicycle almost 2 kilometres to catch the school bus; she was jealous when Rob was given a flash BMX bike that always seemed to go much faster.

Then when she turned twelve, Kelly was sent to Kinross Wolaroi School in Orange. Eric had attended Canberra

Grammar as a boarder for six years, and he and Kim were keen that all three of their children should have a similar opportunity and receive the best possible education. 'We didn't have holidays so they could afford to send all of us. It was pretty tight,' Kelly says.

A co-educational school run by the Uniting Church with about six hundred senior students, Kinross was a very different experience for Kelly who nevertheless thrived. 'The first year, Year 7, was hard but after that I was fine. I absolutely loved it,' she says. 'It was just a beautifully balanced school that didn't focus too much on academics. You did a bit of everything.'

Kinross had a very strict dress code, including a Black Watch tartan kilt that Kelly liked so much she still has hers tucked away in a wardrobe. During a preliminary visit to Orange, Kim took her daughter to the school shop to be fitted out by the principal's wife. Now 175 centimetres tall (5 feet, 10 inches), Kelly was below average height for her age and lightly built. 'Mrs Anderson looked at Kelly and she said, "Oh my goodness, you will have to go in a size 10 children's one." At Dalton school she was taller than all the other kids so we never considered her to be tiny.'

Dangly earrings were not allowed and girls with long hair had to wear it up. The latter wasn't a problem for Kelly. 'Mum cut my bloody hair off just before I went to school. I went from always having pigtails to short and gross. I didn't like it. I said to Mum, "How could you do that to me?" I looked like a boy!'

Naturally outgoing, Kelly soon made friends, particularly with three other boarders, each given a nickname relating to where they came from—Forbes (Lynn), Condo (Tara) and

Brewarrina (Kathleen). During term, the girls would sometimes spend weekends at each other's places, with Kelly making the 230-kilometre trip to Tarcoola about once a term. Still in touch with her friends, she hosted a weekend reunion at Coolong in 2022 as part of a deliberate effort to set aside the isolating experiences of the COVID-19 pandemic. 'We stayed up until five in the morning chatting. It was so nice to see them all. They were all good chicks, and they haven't changed.'

Capable at both sports and study, Kelly threw herself into 'anything and everything' at Kinross, including the school's Australian Army cadet unit. Students were expected to participate from Years 7 to 9, and could then elect to continue with cadet training, which focused on teaching self-reliance and leadership. The training involved weekly afternoon sessions, teaching both theory and practical skills such as bushcraft and field engineering, navigation and orienteering, survival, search and rescue, first aid and catering, abseiling and archery.

At the end of Year 9, Kelly elected to stay involved and eventually became a student cadet under officer, looking after all 150 or so girls in the unit. A high achiever, she also became a prefect and house captain, receiving the Wellwood Shield for Sportsmanship, Leadership and School Spirit. Looking back, Kelly believes going to boarding school and Kinross in particular taught her to 'be independent and go hard'. Importantly, it made her aware of her own potential in leadership and paved the way for the next phase in her life.

On her eighteenth birthday, in January 1993, Kelly walked through the gates of the Australian Defence Force Academy (ADFA) in Canberra. She loved the farm, but Eric and Kim had made it clear they wanted all their children to spend around ten years doing something else before they contemplated coming home, so they could be absolutely certain it was the right choice. 'It's a terrible life if you've been made to do it and you don't like it, and it's hard yakka,' Eric told her.

After considering her options, Kelly decided to join the army. Concerned that her parents still had two children to put through school and couldn't really afford to send her to university, she chose ADFA because it offered a free university education as well as a full-time salary. More importantly, it gave her a chance to explore the leadership potential nurtured at Kinross. 'I saw it as a means to grow and develop and learn about myself, which I did,' Kelly says. 'I was good at cadets and I love leadership—that was my drive.'

Established only eight years before Kelly started there, ADFA combines undergraduate studies at the University of New South Wales with three years of military training for future leaders of the navy, army and air force. To get in, Kelly had to finish high school with good enough marks to be accepted into a university course acceptable to the military, be physically fit and show leadership potential. Successfully negotiating the highly competitive and lengthy application process also involved passing an aptitude test, an interview with a careers coach, a medical assessment, a psychological interview to assess her ability to cope with military life, and another interview where she had to explain why she wanted to enlist.

Kelly was one of more than 2300 people who applied to get into the academy at Campbell in 1993. After 'ticking all the boxes', around three hundred were accepted. The vast majority of them were male—Kelly can't recall the exact number now but her army division of 36 people had only about eight females. 'It was big and very scary but like everything I do in life I went, "Oh well", and got on with it,' Kelly says.

Behind the scenes, her family were not quite as sanguine. Kelly's decision to enlist in the army didn't concern them as such, but they did worry about the prospect of their daughter being sent to serve on the front line, given unrest in the Middle East and recent changes in the Australian military, opening up combat-related roles to women.

While Kelly managed to have plenty of fun, officer cadets had to juggle the normal academic workload of a full-time university student, with military training. She chose to do a Bachelor of Arts degree, majoring in economics and management. She also had to study subjects such as military history, strategy and leadership, and handle a considerable amount of physical training, which she found reasonably easy, given her love of sport and being outdoors. Then there was the discipline of military life, which proved a greater challenge.

'You had to have an immaculate uniform anytime you went anywhere, so there was lots of ironing—it's why I don't iron anything anymore! And I did get into lots of trouble too. Discipline I'm usually really good at, but when I wanted to party and have fun, I was really bad at it,' she confesses. 'For the first year, I'd say I was very well behaved. The second year I started pushing the boundaries a bit and by the third year I was really, really pushing the boundaries. I always passed the academic side, that was never a worry, and I never had to struggle with fitness. It was all pretty easy, that part. But I had lots of stoppages of leave, lots of restrictions of privileges and marching the square in my uniform.'

Kelly graduated in 1995, planning to spend a fourth year at the Royal Military College at Duntroon, but she was diagnosed with asthma and told her future in the army lay not in the field but sitting behind a desk. She hated the idea and left. Meanwhile, her closest army friends went on to complete at least five tours, serving in hot spots such as Afghanistan and Iraq.

Looking back at her time in the army, Kelly sees clearly how it gave her a level of self-awareness she has since put to good use. 'I learnt my strengths and weaknesses really quickly, and then as I moved on after the military I tried to work on my weaknesses and play on my strengths,' she says. 'My strengths are my drive—I just don't ever give up—and I suppose resilience. That was useful moving into rural life because I never let anyone say to me, you can't do that. I just knew I was strong. I could do anything—don't ever tell me I can't because I can. Lots of that was about growing up in the country, and Mum and Dad expecting me to push myself. My worst trait is definitely avoiding conflict. I try to mediate and avoid conflict in most scenarios. It's a bad thing because at times I keep riding something out when I should probably call it.'

After leaving the army Kelly found work with AuctionsPlus, an online livestock buying and selling service, and studied through the University of Western Sydney to gain a Masters in Applied Science. For her thesis she chose to focus on an issue she saw unfolding in her own community—the social impact of Ovine Johne's Disease. The highly infectious wasting disease generated a crisis in the Australian sheep and wool industries in the late 1990s, with the Central and Southern tablelands of New South Wales becoming epicentres.

The disease is caused by a particular type of bacteria infecting an animal's intestines. Sheep pick it up by drinking water or eating pasture contaminated by infected manure. Symptoms can take years to develop, with animals slowly losing condition. At the time there was no way to test live animals to confirm the cause; diagnosis was only possible by autopsy. There is still no cure or treatment, although these days a vaccine is available to reduce the likelihood of death and the amount of bacteria a sheep can potentially shed into the environment.

The first official reported case in Australia was detected near Bathurst in 1980. It's highly likely the disease had been around for some time but it didn't become a major issue until the mid-1990s when it was detected in other parts of New South Wales, most likely spread through the movement of infected sheep. By January 1996, state authorities had identified 80 infected flocks, including prized merino studs that sold rams to other properties. Within twelve months the official number had doubled, with some industry organisations estimating the real number was more like 300.

Fearful of how fast the disease was spreading and the threat it posed, the New South Wales government announced plans in July 1997 to slaughter as many as a million sheep in an attempt to eradicate it. Infected properties were expected to remain destocked for two years before they could be declared disease free and able to bring in replacement flocks.

Meanwhile, other states refused to let sheep from New South Wales pass through their borders unless they had been tested and certified, even though they had infected flocks too.

Critics, and there were many, felt the measures were draconian because too little was known about the real impact of the disease on the viability of the sheep industry, and how to control it. They questioned whether it was realistic to try and eradicate the disease, or even desirable given the financial cost to farmers, the extraordinary mental stress involved and the loss of generations of carefully developed genetics.

Angry about how the issue was being handled and the lack of proper compensation being offered, thousands of New South Wales farmers refused to have their flocks tested or to send their sheep to slaughter. With fears and tempers flaring, one unsuspecting sheep producer from Gunning even reported receiving death threats after selling some wethers that were later found to be infected. 'Johne's devastated this area,' Kelly says. 'People were made to feel like lepers, and there were a couple of families that were pretty much wiped out. They just sold all their sheep and left the farm. It was devastating. Once people tested positive to Johne's it ruined their lives.'

The issue was still causing controversy in 1998 when Kelly became the first woman in the history of Elders Limited to be made a district wool manager. She was offered the position after spending time with the company as part of its graduate program, trying her hand at different roles in different branches of the iconic business, which has been providing services to Australian farmers for more than 180 years.

Aged only 23, Kelly was extremely excited about the opportunity. However, the offer came with a warning. She would be based at Mudgee, where she would have to cope with Phil Jones, who would effectively be her boss, introducing her to clients and mentoring her in what was involved. 'He's been doing it for a very long time, he's awesome but you might find it really hard to get on with him,' she was told.

Originally from Goulburn, Phil is a wool man through and through, although that was not his original plan. He had no intention of following in his father's footsteps and working in the wool trade when he finished school. Instead, he'd been accepted to train as a nurse in Sydney. To earn money and fill in time before his course started, he worked for Elders, wheeling around bales of wool being offered at auction. Then the state manager rang him one day and asked if he wanted to join the business permanently and learn how to sell wool. 'Just give it a go,' he urged. Phil ended up working for Elders for almost 30 years, including a lengthy stint as district wool manager based at Mudgee, the heart of an area well known for producing world-class, superfine fleece.

There is a smile in Kelly's voice when she conveys her first impression of Phil. 'He was an arrogant son of a bitch,' she says. Phil doesn't hesitate to agree. 'I was rude and arrogant, but I thought, "If you can handle this, you can handle anything,"' he explains. 'She was the first female wool rep for Elders so we had to make it work.'

As she does with every challenge, Kelly 'went in hard', much as she had in the army. In her mind it wasn't about proving herself as a woman but proving to herself that she could do the job. As a district wool manager, she was responsible for working with wool producers to prepare their fleece for sale and organise it going to auction. The overwhelming majority of farmer clients supported her appointment and were willing to leave their business in her hands, recognising someone with a genuine passion for sheep and wool, who came from the same background. 'These people were my people. It was just about building relationships with them. I'd say 80 or 90 per cent of clients were amazing. They were happy. Then there was around 10 per cent that were hard-core male, who wanted to go back to working with Phil.'

Along the way there were some unique experiences, including one older client asking her to accompany him to a wedding to keep his business. Kelly politely refused, and kept the business anyway. On another occasion she showed up at a farmhouse, after making an appointment, to be greeted at the door by the client, who was stark naked. 'Would you like a cup of tea?' he asked her, as if there was nothing unusual about it. Kelly calmly accepted the offer, sat down and had a cuppa, talked business and then left.

Despite his 'hard-core' approach, or maybe even because of it, within a week of working with Phil, Kelly had started to fall for him. On his part, Phil had just been through a divorce so he wasn't interested in another relationship, and, besides, they worked together. Some time later, they were at the pub with friends. Phil was sitting at the bar with one of his best mates when Kelly came over, inviting him to join her for a chat. Instead, he suggested she go and talk to another bloke who was also in the bar that night. 'Basically, he tried to hook me up with someone else!'

After Kelly walked away, Phil's mate turned to him and said, 'She actually likes you, Phil.'

'Mate, I've just been divorced. I want no part of this,' Phil replied.

Later that night, Kelly and Phil found themselves standing outside the pub, both looking for taxis. They couldn't find any so Phil offered to walk her home. They were sitting in the lounge room when Kelly took the initiative. 'Kel being Kel, if she wants something she just goes for it. All of a sudden she jumped on my lap and kissed me,' Phil says.

'I really like you,' Kelly told him.

'That's really cool. I'm going home. I've got to feed the dog,' Phil responded and then he just got up and walked out, leaving Kelly feeling like she had made a complete fool of herself.

Kelly didn't give up and about eighteen months after they first met, Phil admitted he had fallen in love. However, he steadfastly refused to stay overnight until she had passed an exam to gain her licence in futures trading. Elders required it so she could advise clients on selling their wool. The exam is challenging—candidates need 80 per cent or more to pass, and Phil didn't want to distract her. 'That's how awful he was,' Kelly jokes. 'He did it to drive me.'

On her first attempt Kelly missed out by 1 per cent. Phil was not happy, and nor was Kelly. 'I don't like failing,' she admits. Two weeks later she sat the exam again, this time achieving 87 per cent.

In the meantime, Elders made Phil redundant. Phil takes some pride in the fact that Kelly was doing her job so well he was no longer needed. Eighteen months later they asked him to come back, and he did for a while, before joining a couple of mates who had gone to work for Australian Wool

Network. After more than 40 years in the wool trade, he still loves it, visiting farms from Cooma to Grenfell, and taking their wool to Sydney every week for auction.

Kelly and Phil married in September 2000 at the tiny Bevendale church of St Thomas. The location was an easy choice for the bride. Generations of her family have celebrated life's milestones in the red-brick Anglican church, which stands on ground carved out from what is now Dowling land, owned by one of Kelly's cousins. Reached via a private gravel laneway leading to the farmhouse, it overlooks open paddocks and a sprawling cemetery where gravestones peek out from tall native grasses. The community has been burying its dead here since the 1870s, when the original station owner donated land for the first timber church. In easy sight of the gate is a row of granite headstones marking the final resting places of Hobby and Una, Loral and Winifred, and Eric's brother, Colin. 'This is where Dad and Mum will be buried, and where I'll be buried,' Kelly says matter-of-factly.

Fewer than 30 people gathered on a chilly spring day for the wedding ceremony. Kelly had never dreamt of a big white wedding. Instead, she wore a stylish after-five dress in deep copper brown when she walked into the church on her father's arm, past the timber font where she was christened. There were no bridal attendants or even an official photographer to capture the moment, but the next day about 120 people gathered in the Tarcoola shearing shed for a big party.

After the wedding, Kelly and Phil made Coolong their home. Kelly had been told by two different army doctors that she couldn't have children, so they restored the main red-brick bungalow with a view to regularly entertaining family and friends instead. Extending the back of the house, they built an enclosed side-porch, and a huge open kitchen, lounge and dining room, with large windows overlooking the garden, and sheep grazing peacefully on Coolong's gentle slopes.

Ahead of her wedding, Kelly finally returned permanently to Coolong to be part of the farm business. Phil was supportive and Eric welcomed her with open arms, involving her in making daily management decisions from the start and handing over the farm cheque book within a year. 'Go and have a crack,' he told her. Always quietly encouraging, he never said no outright or told her something was a bad idea. Instead, he might say: 'I don't know about that. Maybe go and have a think about it.'

Kelly's philosophy is that if you work hard people notice and you gain their respect, but she admits it took time to win some people over to dealing with her and not her father. For years, she pushed herself physically, working long days and not expecting any concessions because she was female. 'It was proving to myself as well as maybe a few doubters,' she says. 'And there's always some male chauvinist pigs that I just stay away from because they just bring you down, and you are never going to change them.'

It's easy to overlook just how extraordinary it was at the time for Eric to step back and let his daughter step up. 'Kel was the first seriously active woman farmer in the district who wasn't a farmer's wife, so she sort of broke the status quo when she came home and started taking such an active role in their operation,' explains Hugh Flynn. The same age as

Luke, he grew up in the district and has known the Dowlings all his life.

'When she started farming, I was probably a bit young to hear any negative talk. I was always just aware of how hard she worked and the fact that she never let the idea that a woman couldn't succeed in farming get in her way. 'Cos that was certainly the prevailing attitude, particularly in traditional sheep-grazing areas like Gunning. It was very traditional and had a huge amount of inertia, so for a woman coming in like that I'd say would have raised a few eyebrows. And from day one, her drive was to do things differently and better . . . It's accepted now that kids when they go home to the family farm, they've gone through uni and they've got fresh ideas and they're going to challenge. It's just what happens now, but for that to happen twenty years ago, it was very unusual, and for a woman to be doing it? That's an even bigger deal.'

Kelly's decision also stood out because at that point she was the only person her age in the district who had returned to farm. Rural communities were so pessimistic about the future of agriculture and the wool industry in particular, that younger people were walking away in droves. In the Gunning area, the community was still reeling from the fallout of Johne's disease and the spectacular collapse of the wool industry in the early 1990s.

The sorry saga that well and truly saw the end of wool as an economic powerhouse in the Australian economy, came about when a reserve price scheme protecting farmers' incomes failed. The scheme was established in the 1970s to stabilise the amount growers received for their wool at auction, after years of low prices. Under the scheme, if wool failed to reach the reserve price, it was purchased by the industry's own Australian Wool Corporation (AWC) and stored until the market improved. With market demand and prices climbing to record levels in 1988, the reserve price increased from 645 cents to 870 cents a kilogram. Responding to these signals, farmers like the Dowlings grew more fibre.

What exactly went wrong and who was to blame is still a point of contention more than 30 years later, but wool production reached an all-time high and global demand crashed. The AWC soon found itself purchasing 70 per cent of the wool clip taken to auction. By the time the floor price scheme was finally suspended in February 1991, the corporation was almost $3 billion in debt and holding a stockpile of 4.7 million bales.

Wool prices remained flat right through the 1990s while the stockpile was gradually sold down and various inquiries tried to work out what had gone wrong and what should be done to turn the industry around. The crisis and ongoing low prices inevitably led to many woolgrowers crossing their merinos with other breeds of sheep to focus on producing prime lambs for meat rather than growing wool. Many others turned to cattle or cropping, or walked away from farming altogether, but traditional fine-wool growers like Eric found it very hard to give up on the industry they loved.

Although he produced a few prime lambs at Tarcoola, Eric decided to stay focused on what he was good at and keep growing wool, doing what he could to reduce costs. He had been gradually building up the flock since the 1970s and moving away from the horned merinos traditionally favoured for fine-wool production, because polled sheep are easier to handle. He continued along this path and for a couple of years tried selling his fleeces unskirted and unclassed to reduce costs at shearing.

By the time Kelly came home the wool industry's prospects finally seemed to be improving. Then came the Millennium Drought. Some parts of south-eastern Australia experienced dry conditions for almost ten years, with southern New South Wales particularly hard hit in the mid-2000s. Drinking water for the sheep wasn't so much of an issue for the Dowlings, because they had access to permanent waterholes and creeks, but the paddocks were bare of feed.

'There were nine definite years of drought here—it was a long, long run. We were feeding sheep every year thinking next year will be better but it wasn't. There would have been quite a few years when we didn't make any money at all,' Kelly says. 'I wouldn't have been able to survive without Phil. I didn't draw anything from the business for many years. I didn't even have my own vehicle. I had the farm truck, an old LandCruiser with no air-conditioning, and Phil's work vehicle.'

Eric and Kelly managed to retain the best performing part of the flock and their breeding stock but during the worst of the drought they were forced to buy in expensive grain and hay. To pay the bills, they took on contract fencing work for government projects in Canberra, leaving Kim and Phil to keep an eye on things. Working off-property five days a week they fenced the new National Arboretum, and erected fences in the Stromlo area and Kowen Forest. The experience brought Eric and Kelly even closer. 'Hanging out with Dad every day was the easiest thing ever,' she says.

On weekends father and daughter returned to Tarcoola and fed sheep. One of the lowest moments in Kelly's farming life came during this period. She was out in the paddock feeding out grain when the starving mob raced between the truck and a trailer it was towing, too frenzied to wait for the vehicle to pull up. Three sheep died after being accidentally run over. 'I remember bawling my eyes out,' she says, then after a subdued pause: 'It's good to reflect on those times. For Luke it's pretty hard to understand those years, not having experienced it.'

Gunning was struggling too, with the drought affecting local small businesses and community morale. One of those who stepped up to help people cope with the enormous financial, physical and mental strain was Melinda Medway. Kelly had known Min since childhood, but the two became close friends after Kelly moved back to Dalton, which had left her feeling socially isolated. 'I cried for months. I was such an extrovert after leaving the army and needing to be around people my age, and I came back to nothing,' she says. Min had taken Kelly under her wing, getting her involved in the Gunning tennis club and introducing her to a new circle of friends to help overcome her sense of loneliness.

Now concerned about the wider community, Min came up with the idea of starting the Merino Cafe, where the meals would be prepared by local farm women who were talented cooks, while local youth worked front of house. The cafe simultaneously generated employment and provided a place for people to gather and talk. 'Mental health was a big issue and we needed something in the community to get some heart back,' she says.

The effort to revitalise the drought-affected community led to Min being named a finalist in the 2008 New South Wales

Woman of the Year Awards. Since then she and her husband, Scott, have restored old Cobb & Co stables just out of town as a function centre that provides even more employment. Min also took on the challenge of pushing for better health services in Gunning. At the time she joined the board of the local community health service, it employed seven staff to provide basic community health services but the town didn't have a doctor. Now the service employs 35 staff and support workers, incorporates a medical practice and brings in a range of allied professionals. 'My doctor is there and so is my physio,' says Kelly.

When the drought also claimed the Gunning Show, Kelly and other younger members of the community stepped in to revive the autumn event. With fond memories of the show as a child and running into the pavilion to see if she had won prizes, Kelly didn't want to see it disappear. She served as president of the show society for twelve years, and then stayed on the committee.

Always looking for a fresh challenge, during these pressure-packed first years back at Coolong, Kelly also accepted a nomination to join a Next Generation working group established by Australian Wool Innovation (AWI). Formed in 2005, the initiative brought together about thirty young Australians, who met biannually to brainstorm ideas that might secure the wool industry's future. Then in 2007 she was appointed as a wool ambassador during celebrations marking the 200th anniversary of the Australian wool trade. At the time she wrote a piece for an industry newsletter which challenged merino breeders: 'I know I am sick of the public/peer perception of the wool industry being negative and backwards.

The industry is innovative and the opportunities are in our grasp—we just have to make them happen . . . The drought has lowered morale but I think that it is our time to show off and celebrate our amazing industry.'

Listening to her talk about sheep and wool, it is clear why she was tapped on the shoulder to champion them. 'I'm a little bit sheep crazy,' she admits. 'I think they're an amazing animal in that they can produce a miracle fibre that is biodegradable and anti-flammable. And then you have the advantage of meat as well. And I've seen them live off nothing. And they like being together—they have this mob instinct. I just find them incredible creatures really.'

During the worst of the drought, around the time she was fencing in Canberra, Kelly also found herself taking on one of life's greatest challenges. Defying earlier prognoses of two doctors, she fell pregnant. 'I got really, really sick. Mum took me to the doctor, thinking I had cancer or something, and he said, "Kelly you're pregnant!" I cried all the way home 'cos I didn't want a baby right then. It was the middle of the drought.'

When they got back to Coolong, it was Kim who broke the news to Phil, too excited to hold back and feeling none of Kelly's reservations. 'You're going to be a dad!' she told him.

'I don't think I can do that,' he replied.

'I still remember it,' Kelly says. 'He was planting photinias in the garden. He slept the night, and then in the morning he rolled over and he said, "Well, maybe we could." Thank God!'

At the age of 30, after a difficult pregnancy and six weeks in hospital, Kelly gave birth to a son. 'It was the best day of our lives,' says Phil, who was with Kelly in the delivery suite.

They christened the baby Ned. Kim was gobsmacked with the choice when Phil rang with the news. 'Don't be bloody stupid, what's his real name?' she scolded. 'It's Ned Dowling Jones,' repeated Phil.

With the drought dragging on, the next few years was a juggling act for Kelly, who found herself pulled between caring for Ned and essential work on the farm. 'I wouldn't send Ned back, but managing a property in drought with a baby was awful, and it put more pressure on Dad, which horrifies me,' Kelly says. 'There was no day care in Gunning at that point, until children turned two, and I went back to work three months after I had him, which is something I feel a little bit guilty about but he didn't miss out because he was around family. Mum was a bit of a legend.'

On days Phil was working away from home, Kim was often the primary carer. That usually involved her heading out onto the farm with Kelly and the baby. 'I breastfed Ned until he was two, so Mum would just tag along,' Kelly says, reflecting that Kim and Ned have a beautiful relationship because they spent so much time together. Kelly tried to keep Ned to specific hours for his daytime naps—9 to 11 in the morning, then 1.30 to 3.30 in the afternoons. 'Wake up, get boob, play, sleep . . . I was a sleep Nazi, but it made it easier for Mum. He always got boob milk because he hated formula so I had to express every day.'

Kim recalls taking Ned up to see his grandfather working in the yards on one particular occasion when the boy was an impressionable three or four years old. Eric normally doesn't swear much, but he was having a bad day. 'That's paddock language. We only use it when we are around sheep and dogs,' Kim told Ned. She realised her mistake a day or so later when Phil let the dogs out and they jumped on his son. 'The language that came out of his mouth. Phil was nearly dying!'

Despite the pressure she was under, Kelly never thought of Ned as a hindrance, taking him everywhere she could. One of the family's favourite photos shows Ned sitting in his bouncer in the Coolong woolshed during shearing, with the shearers lined up alongside at their stands. Another news-making image hanging on the living-room wall captures Kelly holding six-month-old Ned while she greets then Prime Minister John Howard. It was taken in October 2006 when Howard was making a two-day tour of drought-affected areas, organised by local federal member of parliament Alby Schultz, whom Kelly knew quite well.

But there were undoubtedly more than a few occasions over the years when Kelly missed an event in Ned's life she would rather have attended. 'Ned is sixteen now and because of that drive to work I've missed lots, so that would be my only regret. He knows when I'm there, I'm there, but Phil was the one who felt more flexible in his job and made sure he was at everything.'

When Ned turned twelve, he followed in his mother's footsteps and started boarding at Kinross. She had loved her time there, and wanted him to have the same experience, going to a school where boys and girls worked together and grew up learning to understand one another. Like many mothers waving goodbye to their children the first time they left for boarding school, Kelly cried. 'I would drop him off and bawl,' she says. 'It's so unfair, it's the hardest time.'

After going through the experience, she fronted Kim to ask if she had found it difficult too, given Kelly couldn't recall her showing any distress. 'Oh Kelly, we cried all the way home,' Kim admitted.

By comparison, Kelly is fortunate. There were no mobile phones in her day, but she speaks with Ned most evenings before he goes to bed. While he loves the school, he often misses home so it has been a bit of a rollercoaster experience for him, but he comes back for holidays, and for weekends whenever he can.

Kelly cherishes her network of a dozen or so close women friends and most weekends at Coolong revolve around hosting a house full of guests, preparing what Phil describes as 'fancy' dinners, sitting around a fire pit built in the backyard by Eric, or taking them out into the paddock for a picnic. But when Ned is home, 'he becomes the centre of our world,' Kelly says.

A working woolshed is not a quiet place. The sharp barks of working dogs. Bleating sheep. Hundreds of small, hard hooves clattering on timber boards. The loud thrum of a generator. The rhythmic hiss and clang of a sheep handler and drafting gates powered by an air-compressor. These are the sounds that fill the shed at Coolong on a damp May morning, when Kelly and Luke are putting their pre-lambing program into action.

Watching them work is like watching a time-and-motion study in ergonomics. In two eight- to ten-hour days, they will push through 3500 sheep, starting early and stopping just long enough to eat homemade soup for lunch back at the house. Between them, Kelly and Luke carry out multiple tasks in one pass—drenching for intestinal worms, checking feet for signs of bacterial infection, trimming overgrown hooves, and drafting pregnant ewes into small mobs so they can be moved into the paddocks where they will stay until lamb marking in September.

Most of the time the siblings work without any need to talk, which is just as well. To alleviate the monotony during tasks like this, Kelly listens to audiobooks, usually non-fiction, while Luke tunes in to his favourite podcasts. Today, he is responsible for keeping up the supply of sheep with help from his huntaway Stewie, so he is pausing frequently to urge the dog on.

Meanwhile, Kelly is in charge of a brand-new sheep handler. With tens of thousands of animals to treat, it not only saves time but reduces the physical strain on her body. The handler has electric sensors that trigger rubber pads to gently clamp around a sheep's body, safely restraining the animal while she administers an oral worm drench. Before dosing each sheep, she checks the colour of its eartag, which flags what year it was born. Different age groups receive different drenches in a fairly complex approach designed, in part, to reduce the risk of worms developing resistance to any particular treatment.

Working behind his sister, Luke is checking feet and trimming overgrown hooves. Any sign of infection, and Kelly administers a dose of antibiotics. It's been wetter than normal on Coolong for three years, causing more problems than usual, so these treatments are an essential part of managing animal health. 'I actually think I've found the last few years harder than the drought. With all the rain it's been so much more work, so I'm feeling tired,' Kelly admits.

After treatment, at the press of a button, sheep are then released through one of three drafting gates. Animals that have an infection or might need to be checked more frequently are siphoned off so they can be put in a paddock close to the sheep yards. Another small pen catches sheep that don't belong with the pregnant ewes, such as wethers and the odd ram. The pregnant ewes go into the largest pen, where they are counted out by hand and released into the yards, one small mob at a time.

The exact size of the mob depends on the size of the paddock to which they are assigned. Kelly keeps track of all this via an app on her mobile phone, which incorporates a map showing every paddock on Coolong and its grazing capacity. Tap on the paddock and up pops a description of the sheep it's holding, how long they have been grazing there, the acreage and the stocking rate. The app effectively replaces the well-thumbed notebook Eric used to carry in a shirt pocket. It not only saves Kelly dreaded time in the office keying in data, but creates a central record everyone can access easily.

Keeping these records meticulously has become even more critical to managing the enterprise in recent years. In 2021, the Dowlings received accreditation from a global program Kelly believes represents the future for her enterprise and building wool's credentials as a natural, eco-friendly fibre. Known as RWS (short for Responsible Wool Standard), it demonstrates to consumers the care farmers take in producing their wool, especially when it comes to animal welfare and land management. That means no mulesing and taking action to regenerate farmland.

Gaining RWS accreditation is not easy. For woolgrowers, the process starts with developing a comprehensive manual that sets out policies and procedures for every action taken to manage their flocks, harvest the fleece and look after soil health and biodiversity. 'It shows best practice for basically every single job we do on the farm,' says Kelly, thumbing through the thick ringbinder of material she spent weeks putting together.

Farmers then have to keep detailed records documenting their actions. That means, for example, creating an animal register for every shearing, capturing exactly what happened, and documenting every single time someone checks on a mob, why they were checking and what they found. The paperwork for just one shearing at one shed can stretch to six or seven pages. Everyone who comes to work on the farm as a casual or contract worker, from livestock transport drivers moving sheep or collecting wool to the shearing and lamb-marking teams, also has to sign a declaration, saying they are aware of its policies on aspects related to their visit.

To make sure these policies are being implemented and the appropriate records kept, an independent auditor visits the Dowlings every August. Apart from checking the paperwork, he inspects sheep and pays a visit to eight designated monitoring points to assess the impact their farming practices are having on the land. Kelly also has to take photos at these sites every three months so comparisons can be made on aspects such groundcover and native vegetation.

'It's more work, and it's not fun having an auditor come out, I must admit, but I think it's the future in the wool industry,' says Kelly. 'If we are going to market wool with a green image then we have to prove it so the buyer and the end retailer can see we look after our sheep and our farm. When I buy wool jumpers now, I look for the RWS logo, and I like the idea it's traceable all the way back to the farm. I think that's really important.'

Kelly's commitment to RWS is all the more impressive given she hates office work. Sitting at an antique kitchen table in a corner of the main room in the house, she pays bills in the evenings during the week and reconciles accounts at weekends. She only took on the farm's routine bookkeeping about five years ago, when Kim handed over the task, but her training as an economist means she likes being on top of the numbers. She regularly compares them against the budgets she and Luke draw up together, and they drive her management decisions. 'I could be known as a control freak!' she acknowledges, not for the first time.

The Dowlings are already reaping the rewards financially from having RWS accreditation, with buyers paying significant premiums for their superfine, 15-to-16 micron wool since it started carrying the logo. When Phil takes it to auction he is targeting buyers who supply European brands making expensive suits, sports and activewear for high-end markets. Their customers have proven particularly sensitive to environmental issues and campaigns run by animal-welfare groups against the practice of mulesing, which involves removing a strip of skin from the breech and tail of a sheep to prevent flystrike. Blowflies pose a significant health and welfare issue for Australian sheep, so the industry has been investing in research to find alternative solutions.

'I mulesed for fifteen years and then I thought, "If someone was watching this Kelly, it is actually really bad." But with the flies and the big numbers of sheep we run, could we manage without doing it? Then I just thought, "How will we know if we don't have a go?" So we stopped mulesing on one place and realised it wasn't actually that bad. Then the next year nothing got mulesed. It has certainly placed many challenges on our system, which we continue to work on so it is more practical, especially for our shearers. We don't have it right yet, but we will find a way to make it work in the long term.'

The Dowlings now spend more time on preventative measures such as a second crutching to remove urine-stained wool that might attract blowflies. Kelly is also making a point of selecting rams with a plainer breech, which means they have fewer skin folds where blowflies can lay their eggs. She hopes this characteristic will be passed on to their progeny, without sacrificing wool quality. 'We want sheep with a big wide bum so there's fewer wrinkles, but at the same time we don't want to breed the plainest sheep ever, because we would lose our amazing wool.'

Despite the office work involved, Kelly has relished the challenge of gaining RWS accreditation and keeping it. 'It's given me a little bounce again. I had got to the point where I needed a new challenge to stay motivated,' she confesses. 'It took a year to prepare and get everything written up but it's given me a big light at the end of the tunnel because I think it's the right thing for the industry. It isn't for everyone—it's a niche market—but I want to be in it and give myself access to the maximum number of wool buyers. Wool for the future—that's my motto I reckon.'

Aside from the RWS system, the Dowlings objectively measure every animal's performance. They test the micron and weigh the fleece of every sheep, and then index their performance against the rest of the flock. 'So when they are one year old we basically know how productive they're going to be,' Kelly explains. The bottom 25 per cent are sold online or privately, and the top 25 per cent of ewes go into the Dowlings' own merino stud flock. Since the first year she came home, Kelly has been sourcing the best quality semen she can to artificially inseminate (AI) these ewes and breed her own rams. Initially, the focus was on producing finer wool but now that has been achieved the priority is lifting fleece weight.

To take a broader perspective about how the enterprise is travelling, the Dowlings are part of a benchmarking group coordinated by a Wagga-based advisory business. The group brings together fourteen farm businesses from across the Southern Tablelands—half family owned and half corporate. They meet three or four times a year, usually on one of the participant's properties. Eric, Luke and Kelly all try to attend because they value and enjoy the experience so much. 'It's the only day we all take away from the farm together really. We prioritise it,' Kelly says. 'The benefit is hanging out with driven people. When you've had enough of sheep work and you've been doing it all week, you go along and everyone forgets the terrible stuff. People talk about their vision . . . It's pretty amazing.'

When Kelly first came home in 2000, the Dowling enterprise comprised just one farm and part of another. Now they have five farms totalling almost 6680 hectares, including Tarcoola and Coolong; Denbeigh, which they bought back into their branch of the family in 2010; and just over 1000 hectares at Eugowra, about 200 kilometres to the north, which is under the eye of a manager employed to help reduce Kelly's workload. Purchased in 2021, it's added diversity in climate and soil type, and is mostly arable, so they can grow grazing crops such as lucerne, oats and rye, and fatten prime lambs. They have expanded the size of some of the other properties too as neighbouring farms came onto the market.

It takes teams of five or six shearers six solid weeks to shear most of the flock in late spring, excluding Coolong where the shearing happens in March, to spread the workload and the risk of flystrike. In recent times they've also started shearing the Denbeigh wethers every six months, to target a new Italian market looking for RWS wool with a shorter staple. Their regular contractor has been working for them since he was fourteen, and one of his team rents a worker's cottage on the property.

At 74, Eric doesn't do much sheep work these days. Instead, he concentrates mostly on building and maintenance projects, which he loves. He designed the new six-stand shed at Coolong, then led the construction effort with help from a longstanding friend, Des, and Phil who was stuck at home for a while during the pandemic. Eric drew on lessons learnt from building a shed on Tarcoola about 30 years ago, and then another shed with six stands on the property where Luke lives with his wife, Theresa, their two daughters, Penny and Poppy, and baby Robbie.

Luke and Theresa married in 2012. Following the family rule, Luke came back to work on the farm the following year, having spent nine years away from home. After finishing his secondary education at Kinross, he gained a degree in civil engineering and then worked for a mining company in Western Australia. Wanting to be closer to home, he tried working for a construction company in Canberra for a while, but hated it.

A lawyer who specialises in estate planning and farm succession, Theresa says Kelly was quick to make her feel part of the family. Phil and Kelly not only agreed to host the wedding at Coolong, but Theresa and Luke moved in with them the following year and stayed for six months while the Dowlings renovated a nearby farmhouse for them to live in. 'We got on famously,' says Kelly.

The same appears to be true of the siblings. 'He is very respectful of me. He didn't come in and want to change things, but he does question things and force me to think. He hasn't been through a big drought yet, so I'm more conservative. He's go-getting and he pushes me for growth, which is good. We're a nice mix. I hone in on the cost, and he spends!' she adds, only partly joking. 'Dad lived on the smell of an oily rag and I didn't take any money from the farm for twelve years. We've sorta had good years ever since Luke came in, so I have to keep reminding him, "Mate, you've got to be ready for the tough times."'

Theresa admires the patience Kelly shows working with Luke. 'There are things he does that would drive me up the wall, but Kel is very chilled,' she says.

'We don't have any dramas ever, really,' Luke adds during a break in conversation around Coolong's large timber dining table. It's a cold winter night and Eric, Kim, Luke, Theresa and their three children have gathered to share a slow-cooked beef curry. The room is noisy with cross-conversations and laughter as they catch up on news. With the three households less than fifteen minutes apart, they see each other often. Watching the extended Dowling clan together, it is clear this is a close-knit family, who enjoy each other's company. That closeness was a saving grace five years ago when they faced a shared tragedy—the death of Kelly's brother, Rob.

Rob was 38, feeling fit and healthy, when he was diagnosed with incurable, stage-four bowel cancer in March 2016. Not drawn to farming like the rest of the family, he was living in Sydney where he worked as finance director of a computer software company. In defiance of the prognosis, two months later he married Laura, with his two teenage daughters from a previous marriage, Emily and Madison, in attendance.

A month after that, the Gunning community came together to show their support. Held at Min and Scott Medway's old coach stables, the event was organised by Kelly, Luke and Theresa to raise funds for Bowel Cancer Australia and educate people about the disease threatening Rob's life. Kelly was astounded at the response, with the combined generosity of the small community and Rob's friends raising $50,000.

News of the event made the *Canberra Times*, which noted Rob's optimism and the effort his family was making to tackle his diagnosis with 'hope and strength and laughter'. 'Medical opinion, as it stands today, as treatment stands, is that the cancer is probably not curable, but Laura and I are very hopeful of overcoming that at some point,' he told the newspaper, before heading off for his fourth round of chemotherapy.

Reluctant at first to be the centre of attention, Rob came around to thinking that it was important to get people talking about bowel cancer and urged them to seek medical advice if they had even just a hint of a symptom. 'If I hadn't, I might not have had the months I've had or the hope of more,' he said.

His own diagnosis had come as a terrible shock. There was no history of bowel cancer in the family, and the only potential symptom he had experienced was some diarrhoea in the month or two before his diagnosis. It was enough for him to go and see a GP, who was so unconcerned he didn't think it was worth having a colonoscopy, but Rob insisted. The procedure revealed a tumour in his lower bowel. Just a week later when surgeons operated to remove it, they discovered the cancer had metastasised and spread to his peritoneal cavity.

Rob lived for another eighteen months, spending his final weeks at the Chris O'Brien Lifehouse, a not-for-profit cancer hospital in Camperdown. 'He was such a strong little bugger . . . he should have died earlier,' Kelly says. 'For the last three months either Luke or I were there every moment. We sat there every day.'

Luke and Kelly often shared the experience, which helped them both to cope, as did Rob's sometimes macabre sense of humour. One night, towards the end, they were both sleeping in his room when they were disturbed by Rob waking and sitting up in bed. 'Do you think I'm going to die tonight?' he challenged them. 'I'm not! Go back to sleep.'

As Rob's final days unfolded, neighbours dealt with shearing while the family went through the bittersweet experience of welcoming a new life into the world. In late September 2017, Laura gave birth to a baby boy, Sam. 'So we had Laura and this new baby, who are both very much part of the family,' says Kelly, grappling with the complex relationships left behind with Rob's first wife and daughters still very much part of the family too.

Rob died two weeks later. 'None of us could talk about it for a long time,' Kelly admits. In honour of his memory the family placed a granite seat just below her parents' house, overlooking the river where she and Rob played together as children. 'We all go and sit there every now and again for a bit of reflection. Sometimes the sheep come and sit up there too—Rob would find that hilarious.'

Kelly is not sure why, but she also remembers Rob whenever she drives along a certain stretch of road between Coolong and Tarcoola. In her mind, she tells him about her day while passing through. 'I don't know why it happens where it does because the place is of no significance, but it's the spot where I think about him every time.'

Proving the adage that troubles always come in threes, during this terrible period for the Dowlings, Eric was diagnosed with cancer too. In this case, the doctors caught it in the early stages through a piece of good luck, initially presenting itself as misfortune. Eric was cutting firewood with a chainsaw in April 2017 when a tree limb hit him and broke a few ribs. Scans to assess the damage revealed a tumour on one of his kidneys. 'He lost his kidney but it hadn't gone anywhere else,' says Kelly.

Then, ten months after Rob died, Kelly almost died. She was in Sydney about to run her first half marathon to raise money for bowel cancer research when she started experiencing severe abdominal pain. 'You'd better take me to hospital,' she told Phil.

About two years before she'd had half her large intestine removed because of diverticulitis. In the months following the surgery her weight dropped from 96 to 60 kilograms. She'd had a few lesser bouts since then, but after weeks of training she was feeling super fit and ready for the big race. 'I ended up having emergency bowel surgery, and they thought they'd lose me,' Kelly says.

Her condition was so serious that Kim raced to Orange to collect Ned from school and take him to his mother's bedside. An obstruction had cut off blood supply to her small intestine causing strangulation, a condition that can quickly lead to gangrene, peritonitis and death. This time the surgeon had to remove half of Kelly's small intestine. He also had to remove the valve connecting her small and large intestines and make a new one. Phil shakes his head when he recalls his wife's first concern when she woke up after the surgery. 'She was like death warmed up and the first thing she said to the doctor was, "Am I allowed to run tomorrow?" That was his biggest challenge, that doctor, he had to say no to Kel!'

Kelly's recovery took a couple of months, with long days spent lying on the couch at home unable to move. Min came when she could to help shower her, and a cleaner was organised by friends to look after the house. Keen to get back to work, Kelly wasn't the easiest of patients. 'I'd push her back onto the lounge and go, "No, you're lying there a bit longer." She felt like she was letting her father down,' Phil explains.

He and the rest of Kelly's family clearly worry about her and believe that she pushes herself too hard physically. They notice when she picks at her food, or doesn't make time to eat properly, but she is definitely trying. Every morning she gets up early to go for a four-kilometre walk, and two nights a week she goes to a gym in Gunning to box and improve her core body strength. On Saturday mornings, she has a session with a personal trainer in Goulburn, whose business struggled during the pandemic. 'She is a beautiful lady and I love her to death, so that's why I go, to support her business. She works me hard . . . but she knows what drives me. She's been with me through everything.'

As she approaches her fifties, Kelly is getting close to achieving the goal she set for the Dowling enterprise. She wants to hit the thousand-bale mark in wool production. By 2022 they were just eighty bales short. She reckons another 8000 sheep would easily bring it home but that means buying another farm because they are already running at capacity, bearing in mind RWS expectations that the land will be grazed sustainably. 'We have to find the balance,' she says.

Kelly is also keen to complete the jigsaw puzzle of land her branch of the Dowlings own by acquiring a small parcel that sits between two of their properties. It would give them contiguous paddocks, making it easier to manage them and move stock. To reduce the toll on her body, she is looking to employ someone to do more of the physically demanding farm work. 'I'm just finding it harder. I've got terrible discs from lifting sheep and all that sort of stuff. Dad is still doing it but I think it's been a lot harder on my body, as a woman, to be quite honest.'

And she is preparing herself for the next phase of Phil's life, and the changes that will bring. Twelve years her senior, he is fast approaching retirement age. With Ned about to graduate from boarding school, the financial pressures will ease so Phil has the option to wind down gradually and work part-time. He is already responsible for most of the cooking during the week and he does all the gardening, but he doesn't like sitting still so Kelly is encouraging him to think about taking on farm maintenance.

She worries about Eric too and whether he might sometimes feel left out, now that Kelly and Luke are very much in the driving seat. 'But I think he looks at us both and says, "Wow, what a team—my daughter who's been through the shit times and my son who is gung-ho." I think he enjoys it, and he really has walked away. It's a big thing. A lot of farmers wait until they are 80 and their son is 60 and hasn't written a cheque or made a decision yet.'

In the twenty or so years since she came back to Dalton, the scale of the Dowling enterprise has almost quadrupled. 'Eric knew exactly what he had with Kel and he nurtured her, and this is the result,' says Hugh Flynn. 'It's very much a collaboration between Luke and Kel now, but Kel's first among equals. She is one of the best graziers I know. In terms of being a woolgrower she is up there with the best of them, not just through hard work but her technical knowledge and desire to grow the business sustainably and survive . . . I don't think there would ever be something you could put in front of

Kel where she would say, "No, I can't do that, it's too hard." She'd just put her head down and figure it out.'

The tougher question is what Kelly will do next. She is a goal-setter who thrives on meeting challenges and, according to her own assessment, goes 'a little bit haywire on a flat field'. Maintaining the status quo won't satisfy her for long. She recently completed a company director's course, with a view to taking on board positions in rural organisations, and she has been toying with the idea of going into federal politics. An opportunity came up years ago but Ned was only a baby and the timing was all wrong. 'If I do it, I'd want to go the whole way,' she says, picturing herself as agriculture minister.

She also ponders the future for her son, Ned, who at this stage is planning a career off the farm, which doesn't bother Kelly at all. Already helping out on the property, maybe Luke's girls will take it on. Whether it's family or not, Kelly is keen to encourage the next generation of rural women to see a future in agriculture.

Despite the pressures on her time, she has been part of a national mentoring program for girls from the country since 2018. Another way for Kelly to harness her leadership training, the Country to Canberra initiative matches her up with one high-school student at a time so she can mentor them to achieve their academic and life goals. During the formal part of the program, they catch up via the phone or online once a month, using worksheets provided to guide the conversation.

Kelly has enjoyed the experience so much that she has kept in touch with all three girls mentored so far, encouraging them to call if they ever have anything they need to talk about. 'By the end of it you get to know one another really well, and I get to watch them grow up,' Kelly says, with obvious pleasure. Her first mentee even came to work on the farm for a week.

Kelly became part of the program because she really wants young women to believe they can step up and become leaders in agriculture. 'I like the fact that you might only need to encourage someone a little bit or give them a little bit of advice, and it will bring out their strengths. Empowerment is really what it is. And sometimes it's just them being able to ring me and say, "I've had the worst week. I've stuffed up." And I'll tell them, "I've done lots of stupid things and it's going to be okay. That's how you learn."

'They're all amazing. They will send me stuff they've written and I will go, "Oh my God!" I wasn't that smart, or I was partying too hard, but they are nailing it which is kinda cool . . . I suppose all I try to do is help them believe it's okay to fight for something more. Expectation—put it up there and just strive for it. Don't ever accept the second rung, go for the first one.'

After thinking deeply about the unique challenges women might face in agriculture, she says: 'They shouldn't feel that they have to fight because they are a woman. They can just be themselves and it will come, they will be respected, but you have to earn your stripes, and you have to do that in anything you do.'

Kelly credits her own close network of women friends with giving her strength. 'They are my balance, my rocks and what makes it all worthwhile. They are all successful and amazing women, and I love being around them,' she says. From diverse backgrounds and professions, none of them work in the rural sector, which means when they get together twice a year for a girls' weekend, they don't talk farming. 'It gets me away from farming, which is quite healthy, because I do struggle sometimes. When you live and work on the farm, you're in it all the time.'

Despite these occasional struggles, Kelly wouldn't change a thing about her life so far. Even when she is slogging away at a repetitive task, she finds satisfaction. Like the days when she heads into the hills with Eric and Luke to dig up tussocks or do some fencing. 'If we are way out the back of one of the farms, we always take the billy and a hot plate. We start a fire in the morning and have our lunch out there—a sausage and a bit of bread. They are probably some of the best times.'

3

I Am Woman

RUTH ROBINSON, MANNANARIE, SOUTH AUSTRALIA

The gracious hallway running through the Robinson farmhouse is lined with old photos and memorabilia reflecting generations of history. Among the collection is a trophy in the form of a small wooden keg, bound together with stainless-steel hoops. As trophies go it's not that elaborate but it symbolises a seismic shift in thinking towards women at Australia's oldest agricultural college.

The Gramp Hardy Smith Memorial Prize was established in 1939 to perpetuate the memory of three graduates, scions of famous winemaking families who became leaders in their industry. Tom Hardy, Hugo Gramp and Sidney Hill Smith had died tragically the year before in the Kyeema air disaster. They were among eighteen passengers and crew killed when a commercial flight from Adelaide to Melbourne crashed, in what remains one of Australia's worst civil aviation disasters. An appeal raised money for a prestigious annual award for the best student at Roseworthy Agricultural College, judged on the 'all-round characteristics of manliness, sportsmanship and scholarship'.

The wording on the attached plaque had to be modified when Ruth Robinson became the first woman to receive the prize in 1977. That year, Ruth and eight other women also made history as the first female graduates from Roseworthy since the college was established almost a century before. Ruth decided to stay on for another year and expand her qualifications, hoping to then spend some time travelling before going home to Mannanarie in South Australia's Mid North. Instead, a tumultuous series of events saw her take responsibility for running the family farm at the age of 23.

Ruth Robinson always wanted to be a farmer. The younger of two daughters born to David and Nancy Robinson, she was encouraged by her father even though few other young women in that era openly chose a career in agriculture. When the opportunity became available, he actively pushed her towards attending Roseworthy instead of university because he thought it would provide more practical experience. His father, Jim, had taken a similar pragmatic approach to David's own development as a farmer.

An only child, he was just six years old when Jim first put him in charge of the farm's steel-wheeled Fordson tractor. His father stayed close at hand to keep a watchful eye while David harrowed a paddock, but then climbed down off the machine leaving him to operate it alone. 'That night I felt six feet tall,' David wrote. Looking back at that moment, he explained that his father had always shown great confidence in youngsters. 'His philosophy of encouraging children to be self-reliant was surely an admirable one, and one I try to emulate.'

David was the fourth generation of his family to farm the same patch of ground at Mannanarie, a rural district about sixteen kilometres north of Jamestown, which Nancy dubbed romantically the 'Northern Highlands'. Set in a wide valley with fertile soils and rolling hills, the area attracted pastoralists as early as the late 1840s when Mannanarie station was established. Ruth understands the name was derived from an Indigenous word, manangari, used by the Ngadjuri people to describe a native hollyhock used to make string. In translation, it means 'good string or cord'.

The district was marked for closer settlement after the colony's surveyor-general, George Goyder, mapped it safely on the right side of what is famously known as Goyder's Line, during a major drought in 1865. The imaginary line marked the northernmost boundary of land in South Australia considered suitable for cropping. Based on Goyder's observations, it separated land that could be relied upon to receive enough rain most years to grow cereal, from areas more prone to drought and recommended only for grazing. The boundary passes within about 20 kilometres of Ruth's property.

Backed by confidence in Goyder, a few years later new government legislation opened up the Mid North to people who wanted to crop the land, rather than lease it as pastoral runs. Settlers could buy a square mile (almost 260 hectares) on credit instead of having to pay cash up front, creating opportunities for more pioneers to realise their dreams of becoming landholders. When Mannanarie station was subdivided into smaller holdings and offered for selection in 1872, Ruth's great-great-grandfather, Samuel Robinson, snapped some up. He selected two blocks on the first day they became available, expanding his holdings with further purchases over coming years.

After talk emerged of plans to build a new railway through the area, Samuel subdivided one block to create a new township. A total of 55 quarter-acre lots were surveyed, as well as three carriageways with grand-sounding names such as Belgrave Terrace, Grosvenor Street and Ranelagh Avenue. Advertisements proclaimed the benefits of the location adjoining the proposed railway station, at the heart of an immense wheat-growing district with abundant water and excellent building stone.

The first auction in 1877 generated more than £1000 worth of sales and within a few years Mannanarie boasted a pub, school, store and blacksmith, two churches and its own cricket club, but no railway. The proposed line took an alternative route and development stalled. Today there are few signs of Samuel's speculative venture, except a handful of stone buildings and an incongruous sign marking Grosvenor Street—a dusty gravel track passing through mostly empty allotments.

While the township failed to materialise, the farming district of Mannanarie generally prospered. Samuel was joined by two of his brothers, Daniel and Nathaniel, who also took up land in the area. By the 1930s, Samuel's farm was in the hands of Ruth's grandfather Jim, who was more interested in merino sheep than growing cereal.

Ruth's grandmother was another story. Violet Newman was the daughter of a harness-maker turned farmer from Yankalilla. According to Ruth, she was one of four girls, all very strong women. Not long after the end of the Second World War, Vi established a Jersey dairy cattle stud. 'I've always liked Jersey cows, and got a few three years ago for interest,' she told an Adelaide newspaper in 1950 as part of a story celebrating the achievements of women at the Royal Adelaide Show, outside the traditional spheres of cookery and handicraft competitions. By then, the stud was not only winning championship ribbons at country shows, but claiming significant success against the best of the best at the Adelaide event.

Proving her cows were not just excellent in conformation, in 1953 one of Vi's heifers broke a South Australian record. The young cow produced milk containing more than 257 kilograms of butterfat in one lactation period, to become the highest producing two-year-old of any breed officially tested and recorded in the state. This was an important measure in which to succeed, because farmers were paid for the fat content of their milk.

Experts were particularly impressed given the Mannanarie district was not considered ideal dairying country because of its relatively low rainfall, extreme temperatures and limited access to green pasture. Snow is not uncommon, along with frequent severe frosts and chilly winds, while the summers can be hot and dry, with searing northerlies. The average rainfall is only about 425 millimetres, usually concentrated between April and October, and the Robinson farm had no irrigation. Traditionally, farms in the area often kept a cow or two to supply the house, but for many years the Robinson herd was the most northerly being tested for commercial milk production.

Cows became David's passion too when he returned home from attending St Peter's College, one of Adelaide's most prestigious private schools. The stud developed a national reputation under his management, with David later becoming a frequent judge at cattle shows and serving as South Australian president of the Australian Jersey Herd Society.

The cows were milked under a wide lean-to attached to a two-storey stone barn built by Samuel in the early years of establishing the property. Standing just inside the main gate, its solid presence speaks to his aspirations to become a farmer of worth. The main space, where Ruth today stores her wool, is so large it hosted community dances and table-tennis tournaments. Extensive room was also set aside to stable the property's horses, store grain and shear sheep. A small chaffcutter was installed in the upstairs loft to chop up sheafed oaten hay. David later added a hammermill to process grain and removed part of the floor to accommodate an internal silo. A combination of grain and chaff was tipped into a long row of wooden mangers where the cows were fed before every milking.

In the early days, the milk was semi-processed on the farm. A separator was used to extract the cream, which was sent to a local butter factory, and the rest fed to the farm's pig herd. At one stage, the Robinsons even created their own label, DANAR—a combination of the first two letters of David and Nancy's names, with an 'R' for Robinson added to the end. Nancy used it when she packaged fresh Jersey cream and sold it direct to shops in nearby towns.

In later years, the milk was sent unprocessed to the Golden North factory at Laura, which became well known for its prize-winning ice-cream. Initially, David had to deliver it himself, driving around 40 kilometres each way every second day with a load of 10-gallon (45-litre) metal cans. By the time Ruth graduated from Roseworthy, the factory was sending a refrigerated tanker to collect the farm's produce.

While Nancy often fed the calves and took an active interest in the farm, over the years she became increasingly focused on writing. Nancy was the daughter of a well-known Anglican clergyman, the Reverend Cedric William Lyon Noon, who was serving in the Barossa Valley when she was born at Gawler in 1929. She spent most of her young life living in the country towns where he served. A talented scholar and gifted writer who went on to write or edit eight books, Nancy had her first story published in a leading newspaper when she was only twelve. She completed her secondary education at St Peter's Girls' School in Adelaide and then found work in the school broadcasting section of the Australian Broadcasting Commission (ABC), where she started on 'the bottom rung'.

Nancy married David Robinson in 1951. They met through her brother who went to the same college as David, and the wedding was held in the St Peter's College chapel, with Reverend Noon officiating. The newlyweds moved in with David's parents, sharing a substantial stone farm-house built while Jim was fighting overseas during the First World War. Although it lasted until Ruth was about four, the arrangement was far from comfortable. 'The two women were just completely different,' she explains. 'My father and grandfather would have continued to work together quite well on the farm, but the two women [in one house] was a bit much.' The elder Robinsons ended up retiring early to a seaside suburb in Adelaide, and then later a small property in the Adelaide Hills, rarely visiting the farm thereafter.

Part of the tension revolved around Nancy's career. According to Nancy, her mother-in-law disapproved strongly when she learnt Nancy had taken up work as a freelance journalist. Vi thought 'it wasn't quite nice'. Conscious of the trouble it might cause, Nancy initially wrote for newspapers and magazines under various pseudonyms. After realising no-one was covering the area for Adelaide media, she became a correspondent for the afternoon daily, *The News*, and the ABC.

There was no hiding in the early 1970s, when Nancy joined a popular segment on commercial radio hosted by veteran broadcaster Mel Cameron. Called *Mel and the Girls*, the series was broadcast live on 5DN. Nancy co-hosted the session once a month, doing her own research and then driving to Adelaide to interview people as part of a relaxed format that coincided with the earliest days of talkback radio. Nancy also recorded stories for the ABC rural department for many years, sometimes travelling to the studios in Port Pirie, and at others heading out into the middle of a paddock to find a quiet spot and record them at home.

Around the same time Nancy stepped into the limelight with her writing too, when her first book was published under her own name. An in-depth history of Jamestown and district, it was followed by several other local histories and then, in 1974, a groundbreaking book providing advice to women coping with breast surgery. *Sweet Breathes the Breast* was described as the first Australian handbook on mastectomy rehabilitation. It filled such a huge gap internationally that it was eventually translated into 26 languages.

Given that she had no personal experience of breast cancer or surgery, Nancy was often asked why she took on writing such a book. After all, it had not been easy. The research alone involved travelling thousands of miles and making contact with hundreds of patients, as well as scores of medical experts, to gather their knowledge and personal experiences. The catalyst was, in fact, an interview Nancy conducted for her 5DN radio segment, with an Australian businesswoman who'd had a mastectomy and coped remarkably well. Around the same time, she heard about another woman, who had virtually hidden herself away for fourteen years. Intrigued by these contrasting experiences, Nancy set out to learn more. While she found plenty of academic treatises written for medical professionals, she couldn't find anything for women suddenly confronted with losing a breast, about how to prepare for surgery and cope afterwards.

So Nancy wrote the handbook, based on the experiences of as many women as possible. Recognising that she would have to establish a rapport with her readers if they were to act on the advice, she adopted a warm, reassuring tone while explaining the operation and post-recovery phases, and providing practical advice on the physical, emotional and cosmetic challenges involved. Not satisfied with the book alone, in 1974 Nancy also established a self-help association for South Australian mastectomy patients, which provided opportunities for women to meet and share experiences.

In 1976, Nancy and David founded their own small publishing house, Nadjuri Australia. Working on an old kitchen table in the farm office cum sewing room, she took primary responsibility for the business while David continued to focus on the farm, helped by a resident workman. Despite this demarcation and her obvious determination to have a fulfilling career even after marrying, Nancy continued to see herself as a 'traditional woman', who put her husband and family first. 'She was just a loving caring mum,' Ruth explained simply, when asked to describe her.

Ruth had a happy childhood. Born in December 1956, one of her earliest memories was of riding her tricycle up Brighton Road in Adelaide, not on the footpath but in among the traffic, when she was about three. Ruth, her mother and sister were staying at the Noon house, and Ruth had grown tired of waiting to visit Nancy's mother, who was in hospital dying of cancer, so she set off by herself. 'I wondered what the panic was when they came looking for me!'

She also remembers being taken to a speech therapist because her speech was developing too slowly. Adults may have worried, but Ruth wasn't too concerned because her older sister, Mary, acted as an interpreter. That was fine within the family but not so helpful when she started school. Early in her education she was part of a skit performed at the annual Mannanarie Primary School concert, in which she had to say a line about going to get an ambulance. 'I pronounced it *umbalance* and the audience laughed, which I didn't like.'

Ruth attended the one-room school at Mannanarie for the first two years of her education, along with her sister, Mary, who is four years older. The two girls rode their bicycles more than 6 kilometres to join around twenty other children in the classroom, under the tutelage of an unreliable teacher, who often failed to show up on Monday mornings after a weekend binge drinking in Adelaide. Unhappy with this situation, Nancy and David moved their daughters to the primary school in Jamestown. Even though it was expensive, they were then sent to their mother's old school in Adelaide for their secondary education. Mary had completed her leaving exams and joined the workforce by the time Ruth started.

There were no boarding facilities at St Peter's Girls', which was located in the foothills at Stonyfell, so Ruth lodged with an older couple at nearby Maylands. She shared the house with students attending Wattle Park Teachers College, who tended to take the younger girl under their wing. And boarding nearby was a friend of Ruth's from Jamestown, Sarah Pammenter, who was a year ahead at school. 'I would ride my bike to her place every morning, and then we would walk a short distance to catch the school bus. And at weekends we would explore the neighbourhood on our bikes,' Ruth says.

There were very few other country students in her cohort at St Peter's. Her best friend at school was Jane Stapledon, who lived with her parents in Adelaide. Ruth sometimes stayed with Jane's family at weekends, but every four to six weeks she would return home to Mannanarie, travelling by train to Gladstone with other students taking advantage of exeat weekends. 'I didn't ever like Adelaide,' Ruth confesses. 'From Stonyfell you could look out over the city and there was this layer of smog, worse than it is now.'

A gifted scholar, Ruth did well at school despite missing home, taking on extra subjects in her final two years, as well as playing in the school hockey team and joining the debating club. Ruth loved debating, particularly the art of making effective rebuttals. She was also a very keen hockey player, even though it put her in hospital. Playing half back, she broke her nose during one game after she smacked headfirst into a teammate when they both ran for the ball at the same time. 'There was blood everywhere,' recalls Ruth. 'I was staying with Jane for the weekend so her mother took me under her wing and got me to the doctor. Then I had to have the thing reset in hospital. The saving grace was that one of the nurses was from Jamestown.'

As her schooling progressed, Ruth toyed with the idea of becoming a veterinarian. Then she 'lost the plot' after falling sick and missing two matriculation exams. 'I sat supplementary exams but my mates weren't in the room with me and by then the adrenalin had stopped. I did adequately but I didn't really want to be at school anymore. I really just wanted to come home to the farm,' she admits.

Well aware of his daughter's capabilities, David encouraged Ruth to enrol at Roseworthy, making her the first tertiary student in her family. Her exam scores were more than adequate to get into an agricultural science course at university but he believed the college and its Diploma of Agriculture would be more practical, given her determination to become a farmer.

Roseworthy had announced it was opening to girls, but there was still one hurdle. No doubt mindful of the significance of this moment in its history, and the importance of the first female scholars having the ability to thrive, the college insisted each applicant attend an interview with then deputy principal and distinguished horticultural researcher, Milton Spurling. Ruth doesn't recall much about it, except that her parents came too and that he mainly wanted to know why she wanted the opportunity. 'He was a nice bloke. A man of few words and a little stern, but he and I got each other's measure even though the one subject I really didn't like was horticulture and I think he was disappointed that I didn't appreciate it.'

When Ruth Robinson drove through the gates of Roseworthy Agricultural College to begin her studies, she was taking on an entrenched male domain with a longstanding reputation for excellence in agricultural education and research. Established in 1883, for most of its history the college operated as a sub-department of the state government, ultimately under the supervision of the agricultural minister. That all changed in 1973 when a Labor government led by socially progressive premier Don Dunstan passed legislation giving it autonomy as a fully-fledged college of advanced education. One of the clauses in the new Act stipulated that the college could not discriminate against people on the grounds of sex, race, marital status, religious or political beliefs. Then, at the beginning of 1974, free tertiary education was introduced across Australia when Prime Minister Gough Whitlam abolished university and technical college tuition fees.

These significant changes coincided with the departure of longstanding principal Bob Herriot, who had openly declared that girls would be allowed at Roseworthy 'over his dead body'. His replacement held starkly contrasting views. A rural economist with degrees from Sydney and Melbourne universities in agricultural science and commerce, Dr Don Williams had already notched up a distinguished career in agricultural extension—a field that involves educating farmers about the latest scientific research and knowledge, and helping them to apply it.

Given the new title of director, Dr Williams set about transforming Roseworthy into a modern college offering more courses to more students and raising the bar even higher for its research capabilities. The students soon gave him the nickname Bobo, because he had a round cheerful face like the clown on the label of Bobo cordial. 'He was quite determined to bring a new era to Roseworthy, and that included girls,' Ruth says.

Ruth was one of ten young women and about 60 students who started first-year studies at Roseworthy in 1974. Less than half of the girls came from farms; the rest were city dwellers. They included her good friend, Jane, who had decided to enrol, mostly in rebellion against her strict upbringing. 'I was having conflict with my father about what I was going to do when I left school. He wanted me to repeat the year and improve my grades so I could study medicine and I said, “No way! Ruth’s going to Roseworthy, that’s where I want to go.” And that was my ticket to leave home and be my own person,’ Jane recalled in an interview marking the 40th anniversary of their graduation.

Roseworthy was essentially a residential college, with the overwhelming majority of students living on campus. The girls were allocated single rooms in the most modern accommodation block, a two-storey red-brick building known as Block 4. They had access to a small communal kitchenette, and shared a bathroom with cubicles housing three or four showers and toilets, and a urinal—a constant reminder this had previously been an exclusively male domain.

Ruth’s room was ‘quite adequate’, with a built-in desk and bookshelves above, a built-in wardrobe and a window facing north, overlooking the library, some lecture rooms and, further in the distance, the college farm sheds. The room did not have an external lock, but a chain was fitted on the inside of the door to every girl’s room to ensure their privacy and security.

While the girls appreciated this added measure, some of the male students were disgruntled at the special treatment and the fact the young women had been given the best single rooms on the campus, leaving the male students to share rooms or make do with less salubrious accommodation. The media attention that the girls received reinforced the impression the girls were being given special treatment, and so did a welcome dinner hosted exclusively for the young women at the principal’s on-campus home.

A few days later the girls managed to avoid a social faux pas that might have made things worse, after being taken aside by a senior student and warned not to accept invitations to the annual graduation dinner. Ruth and her new colleagues had been invited to partner senior students to this special event. 'First-years don't go to graduation dinners,' the future winemaker explained in a friendly tone. 'That was very good advice,' Ruth says. 'He was looking after us because if we had gone it would have been out of step with the blokes.'

Reflecting on these early challenges, Ruth adds: 'It was very tricky. We were just trying to blend in so there was a bit of a conflict between wanting to prove ourselves, and having to hide our light under a bushel, and I'm still suffering a bit from that. I've never wanted to make a big thing of being a female farmer. I just want to do what I do and I don't expect any special treatment.'

Even though they came from very different backgrounds, the girls soon formed a tight unit. 'We were all pretty strong women. I guess to be the first intake, it wasn't something you did without thinking about it. We really wanted to be there and we wanted to succeed,' Ruth says.

Most of the staff were supportive too, with the exception of an agronomy lecturer who told them at one of their first lectures that they were obviously only at Roseworthy to find husbands. But Ruth recalls this sort of attitude being the exception rather than the rule. She had a particular soft spot for Mr Rowland, the head of the college's agricultural engineering centre, where students learnt skills such as welding. 'He could not have been more helpful. He just bent over backwards. He probably helped all the students but he particularly wanted the girls to succeed.'

Most male students were supportive too, however that didn't mean the girls could avoid something that Roseworthy was well known for—its initiation rituals—although they got off more lightly than students in the past. Roseworthy had a long and sometimes notorious track record when it came to the rites of passage first-years were put through before they were welcomed into the fold. One particularly vicious example in 1930 even triggered a royal commission. Several boys were brutally flogged with cat-o-nine-tails and four students were expelled on orders from the state minister of agriculture.

The incident made headlines again two years later when students at the college went on strike for a day, calling for an inquiry into the principal and the way he was running the college. The government appointed the president of the Industrial Court to investigate events leading up to the strike. He suspected student resentment about the way the flogging incident was handled and attempts to shut down initiation ceremonies might have contributed to the unrest. After hearing evidence, he concluded most of the complaints were without foundation. However, shortly after, the principal resigned and a governing council was charged with looking into 'weaknesses' highlighted during the inquiry and taking steps to remedy them.

Efforts were made over coming decades to stamp out the more repellent initiation practices, but what was referred to as 'orientation' was still in vogue when Ruth arrived. She remembers male students being bullied during sessions when the older students were ostensibly meant to be 'instructing' them on the ways of the college. A fellow student recalled getting very little sleep during her first week when what seemed like a constant roster of senior students shook the girls' doors throughout the night, rattling the security chains. 'There was other stuff that went on with the male first-years that wasn't really very good, but we were quarantined from that a bit,' Ruth admits.

More fun for the athletically inclined was the famous Roseworthy steeplechase, an annual tradition well entrenched by the late 1940s. Initially, it involved first-year students, dubbed 'horses', racing over a four-kilometre cross-country course that might include negotiating ploughed paddocks, fences and vine trellises, before lapping the college oval. Each horse carried the racing colours of an 'owner'—a senior student who purchased the runner at auction. Prices varied according to the level of agility shown jumping over stools and other obstacles in the college common room.

Ruth remembers very little of her own participation in the steeplechase, but Jane recalls being asked to go out on morning runs, which was obviously a ploy to check out her sporting form. She also recalls parading around the 'selling ring' at the auction, where she was purchased for $50, to race for a team of older students who called themselves the Pub Stud. When it came time for the race, a modification was made for the female competitors, which was meant to make it easier. 'The boys ran the whole race but we ran one-third, then got on bikes, and then had to get off with jelly legs and finish the race. [The boys] thought they were doing us a favour but I don't think they really were,' Jane said.

Having survived these rituals, Ruth and her fellow students settled into the routine of learning at Roseworthy, which combined formal lectures, laboratory work and rosters to help look after livestock and carry out various tasks on the college farm. The farm served two purposes: it provided practical experience in cropping, dairying, running a piggery and managing a commercial poultry enterprise; and it gave the college greater opportunities to carry out research trials that helped develop new plant varieties and farming techniques.

Students from a rural background usually found the first year fairly easy. The curriculum focused on basic principles, which also served to bring up to speed the significant number of city students who enrolled. Lectures started at eight o'clock after a cooked breakfast was served in the communal dining room. Each lecture went for 50 minutes, with a 10-minute break between, providing the lecturer finished on time; and then lunch was served back in the dining room. As far as Ruth recalls, at least a few continuous days were set aside in the timetable for lectures, followed by a block of days in a laboratory or out on the farm. Working in the dairy section was her favourite element. 'I probably never had a feel for agronomy. I'm more of an animal person than a plant or machinery person,' she says. 'But I think most of us enjoyed the practical side of it.'

According to Jane, Ruth was a diligent student, who studied hard and stayed focused. 'I was a rebellious sort of person and drank and did all the naughty things you do when you are a rebel, but Ruth didn't,' she says. Weekends were free, except when students were rostered on to help with daily farm chores. The Roseworthy pub was off limits on weeknights but a favourite student haunt on Saturdays.

When she was there, Ruth took particular delight in winding up the blokes by playing Helen Reddy's worldwide 1971 pop hit 'I Am Woman' on the pub jukebox. 'It just resonated,' she says with a grin. Another popular place to gather was Kangaroo Flat, an open space a few kilometres south of the campus. 'We used to go there, armed with booze, stand around a bonfire and drink.'

Then there was hockey, still Ruth's sport of choice. In the first year, eight or nine girls formed Roseworthy's first female hockey team, roping in one or two women from the administration office and lecturers' wives to make up the numbers. Competing against four teams from the Mid North, they won the premiership in their first year, then repeated the feat the following two seasons.

During the week, most of Ruth's peers congregated in the common room after the evening meal, where they could play snooker and watch television. 'A lot of people watched the news at seven o'clock—I usually didn't. I was more likely to fraternise with the older students, and there were a few romances as you might expect,' she adds circumspectly.

Students had to sign in and out of campus, and were discouraged from leaving during the week in what would now be considered incredibly strict provisions for tertiary students. 'I guess we were in a transition phase between the old school and moving into a far more liberated time,' she says. With mobile phones not yet in existence and no landlines to their rooms, students had access to only one phone on campus for personal calls. They were rostered on duty to answer it, and then summon the recipient over a public-address system.

While Ruth enjoyed most aspects of life on campus, the college food was horrible. 'We used to fill up on bread and AJ, apricot jam, because the meals were almost inedible,' she says. The only reliably decent feed of the day was the cooked breakfast. When kitchen staff decided to serve only a light continental breakfast at weekends, Ruth drew up a petition. The decision was reversed. Although she was never a member of the student council, it wasn't the only time that Ruth spoke up on behalf of her classmates. 'I suppose I became a bit of a spokeswoman for the group of girls, and I wasn't afraid to go to the principal if there was something of concern, and he would listen to me.'

As more girls started arriving on campus, the initial group acted as mentors. Under Dr Williams's leadership, Roseworthy expanded rapidly during Ruth's time. By 1976, the number of students had increased by about 70 per cent, to around 270, including seven from Africa and the Middle East who had enrolled in a new graduate diploma for international students.

After three years of study, Ruth graduated with a Diploma of Agriculture, along with all but one of the girls who had dropped out after the first year. Held in the shade of a large marquee pitched on the college oval, the graduation ceremony in March 1977 drew considerable media coverage and scored front cover of the *ROCA Digest*, a periodical published by the Roseworthy Old Collegians Association.

Ruth had something extra to celebrate when she took out the prize for agricultural engineering, and was awarded the Old Students' Cup after coming second overall in her year. A tiny margin separated her and the overall dux of the college, Malcolm Bartholomaeus, who went on to become a highly regarded expert in grain marketing. She was handed an engraved silver teapot by the guest of honour, Lieutenant-Governor Walter Crocker, who grew up on a grazing property near Terowie, only an hour or so from Mannanarie.

Ruth stayed on at Roseworthy for another year to complete a Diploma in Agricultural Technology. 'I think my father probably encouraged me to do that. While he was supportive of me coming home, he wasn't entirely sure it would work, and he thought that I might have to go and get a job in something like the department of agriculture,' she says. 'I specialised in dairying and social science, which was a bit of a mixture of psychology and community dynamics. The course was partly aimed at extension officers, so we had to know how to deal with people.'

In March 1978, Ruth graduated from her diploma course, receiving the Gramp Hardy Smith Trophy in acknowledgement of her well-rounded achievements and contributions to the college. It should have been a joyful occasion, but behind the scenes, hidden even from Jane, Ruth was going through what she still describes as the most terrible experience of her life.

Ruth returned to live at Mannanarie in December 1977. Confident her education would provide a strong foundation for becoming a capable farmer, she was also full of trepidation about the family situation rapidly unfolding at home. 'I would really have liked to work around Australia for a year, but Dad basically told me, "If you want the farm to be here, you'd better come home".'

On Christmas Eve, David revealed to his daughters that he and Nancy were separating. In an awful confluence of events, eleven days later Jim Robinson died. With her sister Mary living and working in Adelaide, Ruth was left sharing the house with her parents while they tried to work their way through a fraught situation. 'It was shear living hell being here with the two of them, both trying to get me on their side.'

With no hope of reconciliation, Nancy moved out in April. 'While it was sad, it was also a relief from the constant tension. Those first months at home were pretty bloody awful frankly,' Ruth says, her voice breaking with emotion even after so many years.

For the next twelve months or so, father and daughter worked alongside each other. At that stage, it was very much a mixed farming enterprise milking about 50 dairy cows, with a piggery of about 30 sows, some sheep and cropping. Mornings were taken up with milking and feeding the pigs, leaving a few spare hours for anything else that needed doing before milking again in the afternoon. Ruth wasn't fond of the six o'clock starts, but it was better than Roseworthy where students had to be on deck at five if they were rostered on duty in the dairy.

Before long David decided to sell the pigs and let the resident workman go. Meanwhile, Ruth took on casual work with the agricultural department in Jamestown, helping to organise appointments for staff visiting farms to test cattle for brucellosis as part of a national eradication program. Mobile phones didn't exist yet, so the calls had to be made in the evenings, once farmers had come inside.

Even while doing this work, Ruth was expected to do most of the cooking and keep the house running, but at least it won the approval of some older local women who had difficulty comprehending that she wanted to be a farmer. 'In general, the men were actually more understanding and supportive than some of the women, because back then most women on farms were not really involved in the actual farming, and they didn't understand that I wanted to be. They would say to me, "Have you got a job yet?" And when I took the job at the department of ag they said, "Oh, I see you've got a job now."'

Then in 1979, Ruth found herself alone at Mannanarie and responsible for managing the farm. As soon as divorce laws allowed, her father married his new partner, Toni Garside. Originally from Victoria, Toni was a passionate Guernsey breeder who had been showing cattle since she was a teenager. She later moved to Western Australia and became president of the local branch of the Australian Jersey Herd Society after adding a small Jersey stud to her enterprise. 'They met showing cattle and then it developed into a bit of a romance. My parents' marriage must have been not entirely wonderful or Dad wouldn't have been attracted to her,' Ruth surmises.

Ruth recalls going over to Western Australia with her father to deliver a couple of cows just before her mother left. She made a return trip for the wedding, and Mary went too. Toni's aged parents were aghast that she was marrying a divorced man with two daughters and refused to attend the ceremony, but Vi flew over, wanting to support her son even though she wasn't entirely happy about the situation either.

With both Toni and David passionate about breeding dairy cows, Toni bought a farm in traditional dairying country, near Murray Bridge on the banks of the River Murray. The newlyweds moved there almost immediately, and David took his Jersey herd with him. 'I could have joined them but I had no interest in the land down there, and I didn't think newly married people needed someone else around, being a third wheel on the bicycle,' Ruth says. 'So, at 23, I found myself here on my own with the cows gone and the pigs gone and the workman gone.'

Confident that she could manage, David left her with a self-replacing merino flock of about eight hundred sheep, which he very much considered 'second-class citizens' compared with his dairy cows, and an irascible old Jersey bull.

Having little choice, Ruth decided to focus on building up the sheep flock and improving its genetics. Many producers at the time were shifting from running horned merinos to polled sheep and trying to grow wool with a finer micron to suit changing market demands. 'I was starting with fairly good body conformity, but broad microns and horned rams, and I wasn't keen on them. They seemed a bit stroppy to me and they'd fight with each other,' she explains.

Her father had bought most of his rams from a next-door neighbour, but Ruth decided to look further afield. For a time, she earnt extra money as a shedhand working for a local shearing team and it opened her eyes to what other producers were doing. 'I could see we needed to fine things down, so I swapped to another neighbour who had a merino stud and I was very selective about what I bought. I often bought the highest price rams at their sales because they were the best rams.' Then, in the 1990s, she started buying rams from the North Ashrose stud at Gulnare, and has been with them ever since. Being the only woman selecting and bidding for sheep didn't discourage her. 'I had probably developed a pretty good water-off-the-duck's-back-type attitude by then,' she says.

The attitude of some local service providers didn't put her off either. Ruth recalls one incident where she had trouble with a new set of hydraulic harrows. They were designed in three sections, with the outer two wings folding up so the equipment would fit through farm gates. When Ruth tried to do this, one wing raised while the other lowered. 'I was adamant it was all plugged in properly, so I rang the dealer and he basically told me off. "You must be doing something wrong, you silly woman!" So, of course, he came out and lo and behold, the same thing happened to him. I'm not sure you could say he apologised but he realised the error of his ways and was more inclined to deal with me on a level-pegging after that.'

Ruth had no such trouble with a Jamestown machinery dealer who supplied a new tractor. Ruth and her father negotiated the purchase of the Chamberlain 4280 together, and then David sent her in to finalise the deal by signing the cheque and passing it over. It was in the very early days of her return to the farm, so Ruth describes it as her 'here I am, I'm serious' moment.

One of the first significant investment decisions Ruth was part of making after David left was building a shearing shed. The Robinsons had previously used a neighbour's shed but now sheep were to be the primary focus, it made sense to have their own. David and Ruth decided on a two-stand design with a raised board. Taking pride of place in front of the stands was one of the first rotating round wool tables in the district, made by a local before commercial versions became widely available. It is still in use today. 'If you are skirting wool on your own, you just keep pulling it towards you,' says Ruth, who is also a qualified wool classer.

With a relatively small number of sheep to look after and no cows to milk, Ruth had time to become actively involved in Rural Youth—an organisation for young farmers which offered the opportunity to socialise with people her own age who had similar interests. She was invited to attend the Jamestown branch by a neighbour, and soon became a regular participant in its social and fundraising activities, and meetings involving guest speakers on various aspects of agriculture. Then there were the popular competitions testing skills in everything from livestock judging, shearing, wool handling, dressmaking and cooking, to debating and public speaking. Branch winners went on to compete at regional level. If successful, it was off to the state finals held to coincide with the Royal Adelaide Show.

Despite it being far from her favourite activity, Ruth even entered a cookery competition, and she won a state competition that involved answering questions on general knowledge, Rural Youth and agriculture. Her prize was spending a morning in the Port Pirie studio with ABC rural broadcaster Ian Doyle, while he was presenting the weekday rural report. Then he took her home for breakfast with his wife, Jane, a well-known Adelaide television news presenter.

In 1980 Ruth took on organising the annual state Rural Youth conference in Loxton, and the following year she was elected state president, becoming only the second woman to hold the honour. 'It was a wonderful training ground for running meetings. I really relished chairing a meeting of 150 delegates and taking points of order, and ruling on things. I just loved it,' she says.

Rural Youth also organised exchanges, with members hosted on farms in other regions and states. Ruth was given the opportunity to go to Western Australia, which proved enlightening in a way that she had definitely not expected. At the time she was engaged, but while she was away she started to have doubts. 'I came home and tried to knuckle down but eventually I realised it wasn't going to work. By that time, my fiancé had worked out things weren't quite right too.'

Recognising she was better off staying single than marrying the wrong person, Ruth ended up living pretty much on her own at Mannanarie for seven years, apart from six months when a young woman who had been transferred to the local bank moved in as a boarder. Her father came home to help with sowing cereal crops, which involved lumping heavy bags of seed and fertiliser, but as time rolled on David found it increasingly difficult to get away from the dairy farm. Always practical, Ruth proposed bringing in sharefarmers to handle the cropping program.

Ruth had been home for about six years when she met John Voumard. A lawyer working in Port Pirie, John spent much of his young life at Naracoorte where his father was a stock agent and auctioneer for Elders. Brian Voumard started out as an office boy at the company's Streaky Bay branch on Eyre Peninsula and ended up territory manager of the insurance section. In between, he was constantly on the move as he worked his way up the ranks, but the state's south-east is where John finished his high-school education, apart from a year in the Philippines as a Rotary exchange student.

John spent six years at university in Adelaide gaining a law degree and diploma in legal practice, and making a start on an arts degree, before moving to the Mid North. At the time he met Ruth, he was boarding with a local bank manager, who was a friend of his parents, and still trying to settle into his first full-time job in a town he didn't know. In fact he jokingly recalls asking his father where Port Pirie was when the position came up. He took the job because he was only being short-listed for 'crash bang' law firms in Adelaide, where he would have been stuck chasing uninsured losses from accidents. 'I was well and truly over it and anxious for a real job,' he says.

Ruth and John first encountered each other in the Crystal Brook pub sometime in 1983, at a dinner meeting of the Rocky River Young Liberals. Interested in politics, both had become members of the organisation as a way of meeting like-minded people of their own age. The branch covered the state electorate of Rocky River, which incorporated both Port Pirie and Jamestown until it was abolished in 1985. Members met regularly over dinners held in various locations.

Neither recalls anything remarkable happening that evening, but John's attention was definitely engaged. A couple of years younger than Ruth, he found her more 'switched on and intelligent' than anyone else he had met so far during his brief time at Port Pirie. Ruth began to take more notice of John the following year when they were thrown together in a series of debates run by the Young Liberals. They were part of a team that took on other branches from across the state, with considerable success. 'Then the debates ran out and if we were going to continue to see each other, we had to do something about it,' Ruth says.

Their first official date was lunch with some friends of John's at a Clare Valley winery. By then he was based in Clare, working for the same law firm. Ruth had already turned down an invitation to accompany him to a wedding because she had a hockey game, but this time she said yes, even though she was uncertain of the status of their relationship. 'I felt a little uneasy because his friends were obviously looking at me thinking, "John's got a girlfriend", and I wasn't really sure at that point whether I was his girlfriend. But I definitely was later that day!'

For their next date, John invited Ruth to accompany him to a mid-week opera performance in Adelaide. Ruth had never been to the opera before but she accepted the invitation and bought a new dress. Neither recall what was performed, however it was enough for Ruth to decide she didn't particularly like opera. She did like John. After a 'whirlwind' romance of just a few months, they became engaged and four months later, in April 1985, they married. 'John's father was aghast,' Ruth recalls.

Apart from how quickly it happened, an issue for John's parents was that they were Catholic and Ruth was a Protestant at a time when 'mixed' marriages were not that common. Ruth insisted on being married at the Anglican church in Jamestown, which caused some tension. However, arrangements were made for a Catholic priest to co-officiate with a relation of Ruth's, who was an Anglican priest. The small church was packed for the ceremony, which took place on a warm autumn morning. To avoid having to work out where to seat Ruth's divorced parents, they again defied convention and had a stand-up reception at the football clubrooms. Then the newlyweds drove to Melbourne where they boarded a flight to Tasmania after a luxurious night at the Windsor

Hotel, given to them as a wedding present. Honeymoon over, they came back to live at Mannanarie.

When he decided to marry Ruth, John realised he was also making a commitment to the farm. 'Farms aren't portable. I knew that if Ruth was the one, then it had to be here,' he says. So he negotiated a transfer to his legal firm's Jamestown office, where he was given a desk in the kitchen. The practice shared rooms with a rural supply business and no other space was available. He didn't even have his own phone.

By no means a farmer, in his spare time at home John took on the gardening. One of the first tasks he and Ruth tackled together was putting up a new fence, to replace the hedges Ruth had removed to open up views of the surrounding countryside. The fence panels were ready-made but posts had to be put in to support them. 'That was fairly interesting. They had to be in the right place or it wouldn't work,' Ruth says before adding carefully: 'That probably proved John wasn't really a technician, but we got the fence up anyway.'

John may not be especially handy, but Ruth doesn't know how she would have coped if he hadn't been there when tragedy struck a year later. In the early hours of 5 May 1986, her father died after an accident on the farm at Woods Point. At around four o'clock in the afternoon of the day before, he was using oxyacetylene equipment to cut open a 44-gallon drum when it exploded. The drum contained residual fuel—a tiny amount but more than enough to trigger disaster.

Suffering horrendous third-degree burns, David was rushed to the Royal Adelaide Hospital. Ruth and John were out for the day so they didn't learn of the accident until that evening when they found a message on their answering machine.

'We got there in time to see him but it just didn't look like Dad. He was so swollen up and he couldn't talk. He attempted to say something but I have no idea what it was,' Ruth says, fighting back tears. 'I was just numb with shock, but at least John was there to be with me. He was a rock.'

Ruth had formally taken over the farm business in 1984, and was in the process of purchasing one of the blocks, but David left the entire property to her in his will. Meanwhile, a devastated Toni kept the farm at Woods Point and took on running the Jersey and Guernsey stud herds without him.

During the next few years Ruth was grateful for occasional advice provided by her shearing contractor, Keith Woidt, and an old neighbour, Hillary Baynes. Keith was especially supportive when Ruth told him she was pregnant with her first child, and they needed to move the shearing back so it didn't clash with the baby's arrival. Apart from bringing up sheep and working as a shedhand, Ruth usually provided his shearing team with a hot lunch, and morning and afternoon smoko. But that year, Keith's wife, Yvonne, said, 'Don't worry about the food, I'll just bring it all out.'

The safe arrival of their first child, Catherine, in August 1988 was both a joyous moment and a great relief for Ruth and John. Ruth had experienced several early miscarriages, with her doctor claiming she was too old (at 32) for child-birth and really should have had children sooner. His bluntly expressed opinion upset Ruth, even though it reflected the views of the day, with most women having their first child at a much younger age. Concerned this baby might also be lost, she did her best to follow the doctor's advice after he ordered bed rest until the fourteen-week mark. Ruth kept herself occupied editing a friend's doctoral thesis, while John took charge of feeding the sheep. Carefully following daily instructions issued by his wife, he headed out in the winter dark before going to the office.

Even after the first trimester, the doctor was so concerned he banned Ruth from travelling any great distance during the rest of her pregnancy. During this time, she missed two friends' weddings and could not be there for her friend Jane, when her husband was killed in a car accident. Then, the doctor became worried the baby had stopped growing and ordered Ruth into hospital at Jamestown. A fortnight later Catherine was born by caesarean section. Ruth recalls returning to the farm with the baby, a little panicked about coping. 'We didn't know what the hell we were doing but you muddle through,' she says.

Ruth had to go into hospital one or two weeks before their second child, Sarah, was born almost two years later, also by caesarean. Nancy came back to Mannanarie to help look after Catherine in her mother's absence. After that, Ruth was mostly on her own during the day, juggling the children and managing the farm, as well as the bookwork for John's own legal practice, which he had set up at Peterborough before Catherine was born. 'Only the essentials got done and even those sometimes got done later than they should have been, but looking back it was only a relatively short time in the general scheme of things,' Ruth says.

In some ways, Ruth reckons being a self-employed farmer made motherhood easier. From the time they were babies, she took her children out on the farm, buying a dual-cab ute after Sarah was born and fitting it out with child safety seats. Catherine has a vivid memory of having a fabulous time playing in mud and looking for worms while her mother was fixing a leaking water pipe. 'We were very lucky to be part of her day-to-day work,' she says.

Ruth took the children to ram sales too, earning frowns from some when they reached out from their pushers to pat the sheep. On the one day a week she worked in the legal practice, the children were looked after by a friend who offered family day care. When they became fractious back at home, Ruth put them on her hip and took them out into the yard to play with the dogs, which soon changed their mood.

At shearing time, her mother-in-law, Anne, would come to help look after the children and do the cooking. In more recent years her sister, Mary, has helped too. Apart from a few seasons when one team liked to bring their own food, Ruth has always provided three meals a day, including a hot lunch served at the farmhouse table, and home-baked cakes, biscuits and sandwiches for smoko. It takes considerable planning, given it's also her job to keep up the supply of sheep in the shearing shed, and help do the wool classing. For the two weeks or so it takes to shear about 2300 sheep, she is effectively pulled in three directions. 'I'm mustering ready for the next day, I'm in the shed helping with wool preparation, and I've always got my mind on the next meal. Or I'll be conscious a wool bale is about full and that a shearer's pen is almost empty,' Ruth explains.

As they got older and started going to school, Catherine and Sarah began to realise that having a mother, rather than a father, who was the primary farmer, was far from the norm. 'I was very aware of it growing up,' Catherine says. 'It was a novelty for that fairly conservative country town, and it was also a novelty that she kept her maiden name when she married and she was Ms Robinson and not Mrs Voumard. I now have a lot of respect for the fact that she was known in her profession before she met Dad and decided to keep her name . . . but I don't think Mum would ever describe herself as a feminist. She just did what she loved and what she wanted to do. While always supportive of women's causes, she's always been a true believer of the right person for the job and anyone can do what they want to regardless of their gender.'

The girls were still in primary school when Ruth became heavily involved in the South Australian Farmers Federation (SAFF), the main lobby group for the state's farmers and another predominantly male domain. 'I guess I was generally interested in agripolitics and having a voice,' she says. 'The blokes were willing to take me at face value and, again, I had that confidence from being at Roseworthy. I knew I was no fool, but I wouldn't have gone in there and held the floor and wanted more rights than anyone else,' she emphasises. 'No-one likes pushy people.'

After serving as vice president, Ruth was elected president of the Jamestown branch in 1994, and zone chairman the following year when she also began representing her region on the state board. In 1999, she became the first woman in the organisation's history to hold the position of senior vice president. During this period, she represented SAFF on the National Farmers' Federation Council. She also spent seven years coordinating a future leaders' program.

These commitments took Ruth away from home for a substantial amount of time, often with early starts to make full-day board meetings in Adelaide. It was also a challenging time in the organisation's history, with declining membership and finances, and a difficult 'situation' emerging between two senior figures that Ruth found herself managing but unable to talk about, because it was deemed highly confidential.

Ambitious to capitalise on her experience and achieve more, Ruth took the next logical step and in 2002 stood for election as state president. With reasonable expectations of winning, she and John purchased a unit in Adelaide that would give her somewhere to stay so she was better able to fulfil her duties. But it wasn't to be. A popular challenger with no idea what she had been dealing with, stood against her. After a fiercely fought campaign, he won.

Losing the election hit Ruth hard. Still trying to take it in, she had to front up after the vote was taken at the annual state conference and chair a business session. 'I saw all that out and I put on a brave face for the dinner that night, but I didn't go back the next day,' she confesses. While she remains disappointed that she wasn't successful, Ruth has become more philosophical about the loss over time. 'In the years after that I didn't really know whether to be sorry that I'd missed out, or glad I didn't get it because it would have been a very large time commitment. And I think, in general, they probably weren't ready for a woman at that stage.'

Instead, Ruth threw her energies into serving on the board of the Royal Automobile Association of South Australia (RAA), a prestigious position she held for fourteen years. 'They needed a woman and someone from the country so I ticked both boxes,' she says modestly. She has served on numerous other local, regional and state committees too, from chairing a regional health board to sitting on a committee advising the agricultural minister on regulatory issues relating to the sheep industry, and presiding over the Stuart electorate committee for the Liberal party.

It's a sunny autumn Sunday morning at Mannanarie. John has been out in the garden, planting some vegie seedlings. They were a gift from tenants who have recently moved into the original Mannanarie station homestead, which John and Ruth purchased in 2007 together with more land. Most Sundays, John and Ruth go their separate ways to attend church; John remains a committed Catholic, and Ruth has become a lay preacher in Jamestown's Anglican church. She even studied theology for two or three years to prepare for the role.

Ruth's first taste of religious life came as a child when she and Mary would sit at the kitchen table tackling what was known as Mailbag Sunday School. Packages of books and activity sheets were mailed out to rural homes as part of a service provided by the diocese for families who could not attend church regularly, or where there was no local Sunday school. At that time, the Robinsons attended services at the nearby Yongala church, but they were only held once a month. Ruth also suspects her mother felt under pressure to make sure they had at least a rudimentary knowledge of the Bible, given she was a clergyman's daughter. 'I can remember colouring in pictures, and there would be a little story, and I think as we got older we had to write something and then send it back.'

Religion was a daily part of life at St Peter's Girls' too, given it is an Anglican college. 'We had prayers every morning in the chapel, and scripture lessons once a week. And wherever I was boarding, I'd find out where the local church was and rock up every so often. It was partly a social thing I suppose,' Ruth admits. Even at Roseworthy, she was involved in starting a Christian club. 'So I think there has always been something there for me, and then I got involved in the local church when I came home.'

Ruth ending up sitting on the diocesan council for about 25 years, and represented the diocesan laity at the national general synod for more than ten years. Then a female priest serving the Jamestown parish encouraged her to become a pastoral assistant, which involved helping during the service. He then suggested that she should consider preaching when a priest was not available. Ruth was a confident public speaker, but she felt very strongly that she couldn't deliver a sermon without knowing more about the testaments. Over a period of two or three years, she studied with the Adelaide College of Divinity, which offered intensive weekend courses, and online with Charles Sturt University.

Ruth enjoyed the experience, but could only tackle one subject at a time because of her responsibilities managing the farm. Inevitably, assignments ended up being finished late at night, under pressure, just before they were due. Having covered the core subjects, Ruth set the studies aside and now uses what she learnt to lead a service at her church every second month. While she may spend a few days thinking about what she intends to say, the sermon is usually written the day before, sitting at the desk in her messy office. 'I quite enjoy it. I see my role as helping to interpret for the congregation the readings for the day, and bringing it down to what does it mean for us.'

Occasionally she also preaches at one of the seven other churches in the area. Perhaps her greatest test came a few years ago when Jamestown hosted the annual synod meeting of the Diocese of Willochra, a massive district which covers around 90 per cent of the state. She was asked to preach at the Sunday service and base her sermon around this reading from the Bible: '. . . it is easier for a camel to go through the eye of a needle than for someone who is rich to enter the kingdom of God.' After pondering the challenge for some time, it suddenly came to her that she could make use of the church's Christmas nativity set. It included a relatively large camel. During the service, Ruth gave the camel to a parishioner and stood next to him holding the largest sewing needle that she had been able to find. 'Now Maurie, put the camel through the needle,' she told him. Everyone laughed as Maurie valiantly tried to fulfil her request. 'I think people really remembered it, and I got a buzz out of it. Yes, my knees were knocking but I think if any public address is going to be valuable, you should feel a little bit nervous.'

Ruth may enjoy her role as a lay preacher, but she has never thought about taking it a step further and joining the priesthood, even though this path is now open to women in her diocese. 'I don't feel called to become clergy. My sense of vocation is serving the local community and helping interpret the Bible. I don't have any great sense of needing to be a priest. Then there's the fact that John is a Catholic and a little askance at the idea of women clergy. He's quite happy for me to do what I'm doing but I think it would be a step too far.'

Ruth admits to experiencing times of doubt when it comes to believing in God. Surprisingly, it was studying science at Roseworthy that reinforced her faith. 'I realised the series of chemical reactions that happen in photosynthesis in a plant are virtually the exact opposite to what happens in animals. It's called the Krebs Cycle. It's not simple. It's a series of chemical reactions and I thought, "This can't have just happened. There must have been some architect behind it."'

It is another scientific puzzle that is occupying Ruth's mind when she drives into a small paddock not far from the house. Ruth and John are both skipping church this particular Sunday to drive to the Limestone Coast where Sarah works with the National Parks and Wildlife Service. They will be away for a couple of days and there is one last mob of sheep she wants to feed before they go.

Providing supplementary feed to her livestock has become more onerous than usual in recent weeks because Ruth has volunteered to be part of a research project relating to animal health. Scientists from the South Australian Research and Development Institute are monitoring 400 of her flock to explore the impact of giving pregnant ewes melatonin supplements when they are carrying twins. Initial trials suggest increased levels of the naturally occurring hormone improve the survival rates of twin lambs by 10 per cent.

Manangari is one of eleven farms in South Australia where the research team is testing whether this response can be replicated in commercial-sized flocks. Researchers came to the property to treat the sheep about 90 days into their pregnancy, after scanning confirmed a suitable number were carrying twins. The ewes were divided into four separate mobs.

Two were control groups and won't receive any treatment, while the other ewes were administered melatonin via an implant placed just behind their ears, which allows slow-released doses to pass from the mother to the lambs in utero via the placenta.

The mobs have to be kept in separate paddocks. On top of the rest of her flock, this means Ruth has seven different mobs and a total of about 1500 sheep to manage during the autumn and early winter months when grass is in short supply and they need to be fed supplementary hay and grain every few days. She is willing to make the effort because of the potential benefit to her own sheep as well as countless other merino enterprises across the country.

Lamb deaths cost the sheep industry millions of dollars a year. Most losses occur within the first three days, with around 70 per cent caused by complications during labour. All lambs experience oxygen deprivation, or hypoxia, during birth but the second lamb is at far greater risk because it spends longer in the birth canal, potentially damaging its brain, nervous system and vital organs, reducing its vigour and ability to interact with its mother. Melatonin seems to protect the lambs from hypoxia, and reduce these risks significantly.

Ruth already recognises the importance of nutrition in building up the strength of her pregnant ewes. She gives them lime to boost their calcium levels, which helps prevent milk fever and gives the lambs better bone structure. About a month before lambing, magnesium is added to their feed to improve their muscle tone, so they can push the lambs out more easily, and get up faster after giving birth.

Ruth's level of care reflects her first love in farming, which has always been working with animals. When things go wrong and sheep die, she feels the loss deeply. The 1990s were tough because of dry seasons, soaring interest rates and the collapse of the wool market. She was forced to sell sheep for ridiculously low prices, but she didn't have to shoot any like many other farmers. 'That would have been soul destroying,' she says.

Ruth still blames herself for the night she lost 192 freshly shorn lambs when a cold-snap hit. The weather was mild when she moved them into a lucerne paddock in the bottom of the valley. One night soon after, she and John went to visit Mary at Brinkworth. 'We were tired when we got home, and I wasn't attuned to the fact the weather had changed,' she says. 'I found them in the morning.'

Ruth managed to revive most of the mob of about 975 lambs, but many others were already dead or comatose and beyond saving. She had to borrow a front-end loader and bury them with help from neighbours. 'It was hard to cope with, it really was. These sheep died because I should have shifted them earlier. I was responsible.'

Ruth says it is one of the most challenging moments she has faced as a farmer, along with another terrible incident about five years ago when a driver ploughed into some sheep when she was moving them across a main road. He hadn't noticed the warning signs she'd put out, or the flashing light on top of her ute. The driver was more worried about his dented bumper bar than her sheep and went on his way, leaving a furious Ruth and a passing motorist to round up the scattered mob and deal with the dead and maimed. Several died instantly. Ruth had to shoot a few others that were in pain and too injured to save.

Catherine says that while her mother understands the necessity for taking such action, her heart is always with the sheep at moments like this. 'She just loves animals and loves to see the progress she has made in breeding them and seeing an improvement in the wool.'

Ruth is visiting her favourite place on the farm. Most days the rocky ridge gives her a clear view of the entire Mannanarie district and its surrounding hills. The seasonal break hasn't come yet even though it's now May. The paddocks remain brown, scudding clouds casting dark shadows as they roll across the open valley. The sun is out but a wind has sprung up, fierce and cold, so Ruth has swapped her baseball cap and sun protector for a navy wool beanie.

The viewpoint enables Ruth to take in at a glance the 945 hectares she now farms. Spread across three separate holdings, they include the original Robinson farm of around 510 hectares, which she renamed Manangari; a block of 138 hectares across the road purchased in 1977; and another 297 adjoining hectares, which John and Ruth bought in 2007. They seized the opportunity to make this last purchase when the remaining Mannanarie station block was further subdivided and put up for auction. They bid successfully for two sections, incorporating the original station homestead, built around 1847, which they restored and then rented out.

From her perch on the range, Ruth can also clearly see the outcome of years of her own work planting shelter belts, with advice on species selection from Wirrabara vegetation consultant, Anne Brown. Her grandmother planted trees for shade and so did her father, but they were mostly single plantings or rows placed along roadside fences to help define the property's boundary. By comparison, Ruth has established wide multi-species corridors along internal fence lines to protect her sheep from the most severe weather conditions, while also encouraging native birdlife.

The windbreaks are up to 700 metres long, with five staggered rows of trees and shrubs indigenous to the area, supplied as seedlings by Trees for Life. The middle and tallest row is a local blue gum. On either side are casuarinas and mallee box, which generally grow slightly shorter and have a wide canopy, and the outer rows are shrubs. This combination is slightly permeable to the wind, slowing it down without creating turbulence, unlike a single row of dense plantings.

Getting the windbreaks established can be a heartbreaking exercise, given the hard frosts in winter and the hot, dry summers. Choosing what she hopes is the lesser of two evils, Ruth plants her seedlings in July. Her theory is that the tough ones will survive and won't need as much watering over summer because they have had longer to establish.

Behind Ruth, the hillside is carved with shallow swales of contour banks created by her father to reduce erosion. It's much less of a problem these days, given enormous changes in the way farmers in the area now grow their crops. Many use no-till or minimum tillage practices, sowing in a single pass with little disruption to the topsoil. When Ruth was young, paddocks were cultivated after spring rains ready for sowing the following autumn. 'Then every time it rained and the weeds grew you'd cultivate it again, and then in autumn you'd harrow it to fine it down, so you'd work the ground probably six or seven times before putting the seed in,' she says. This approach left the topsoil exposed, prone to blowing away in dry months or being washed into creeks after heavy rains.

To the south of where Ruth is standing lies flat-topped Mount Lock, the area's highest point at around 730 metres above sea level. Behind her are the towering white turbines of the Hornsdale Wind Farm. More than 90 are spread across 7500 hectares of privately held land, none of it owned by Ruth. She and John objected to the development, which was granted initial planning approval in 2012. Her neighbours receive regular payments for allowing the turbines to be built on their land but the proposed sites on the Robinson farm were within 800 metres of the house. Ruth and John were unhappy about the proximity and potential noise issues, the disruption during construction, and the way access roads would carve up paddocks and make them difficult to manage.

While Ruth does not appreciate the turbines, the environmental studies that had to be carried out before the wind farm was approved led to a pleasant surprise. Ruth learnt that her paddocks were home to an endangered species—the pygmy bluetongue lizard. Thought to be extinct until they were rediscovered near Burra in 1992, the species had been recorded in only a few locations on the Adelaide Plains, with very few specimens collected and only one first-hand report on its ecology, published in 1863. Scientists found specimens on land neighbouring Ruth's properties during an environmental survey for the wind farm. Learning of the discovery, Greening Australia decided it might be worth searching adjacent properties and approached Ruth.

Back in the ute, Ruth heads to the paddock which she describes as 'pygmy bluetongue central'. Let off the chain, Ruth's young and not always obedient kelpie, Scooter, takes off after a couple of kangaroos. Her older dog, Izzy, is happy to stay on the back of the ute, nose into the wind as Ruth drives up a hill and over the other side, towards a seemingly unremarkable patch of lower ground. The paddock is predominantly native grasses and has remained uncultivated for many years. That makes it the perfect habitat for the lizards, which live in abandoned holes made by trapdoor and wolf spiders. 'It just looks like grassland, doesn't it,' Ruth says.

The lizards only grow to about 15 centimetres long. Despite their name, they don't have blue tongues, but their jaw muscles, short tail and legs give them a similar appearance to larger bluetongue species. 'You don't see them because they are very shy little creatures,' Ruth explains, crouching down among the grass searching for likely dwellings. 'They can't dig their own holes, so they move into a hole where a spider has burrowed. You will only find them where those spiders have been able to create holes, which counts out any paddocks that have been cropped. Cultivation destroys the soil structure, so when a spider tries to build a hole it just collapses.'

The lizards spend most of their lives in the ground, poking their heads out to catch insects. The best time to spot them is during mating season in the spring—the only time they move any distance from their homes. This explains why Ruth had no idea of the lizards' existence until the excited researchers from Greening Australia reported back to her. Further investigation by Flinders University has since discovered her lizards are 'bolder' than other populations, making them easier to catch. Concerned about the future impact of climate change on the species, some of the captured lizards have been relocated to a farm further south as part of a major research project led by the university. The $400,000 project is generating knowledge about how to re-establish populations of lizards in cooler areas if climate change makes their current habitat unviable.

Ruth is fascinated by the work and enjoys being kept informed. It's indicative of her growing interest in native flora and fauna, including encouraging native pastures such as wallaby grass, spear grass, and even the endemic iron grass, which isn't palatable to sheep but provides useful shelter for small lambs. 'I strongly believe we should have more native pastures because they don't need high rainfall to survive,' she says.

Her daughter Sarah is enthusiastic about native grasses too. It is something she plans to explore more thoroughly when she eventually takes over Manangari. While Catherine has followed in her father's footsteps and become a lawyer, Sarah aspires to emulate her mother. When the day comes, she will bring a different viewpoint to the enterprise, drawing on a degree in environmental policy and management, and steadily accumulating experience working for landscape boards and the state environment department. Sarah chose this career path because she wanted the opportunity to learn more about environmental management and spend time working outdoors. Just like her mother, she prefers living on a farm to city life, so she has looked for regional positions that often involved working with farmers.

Aware of her mother's interest in encouraging native flora and fauna, Sarah thinks there is potential to take this even further, perhaps planting native grasses and reaping the seeds as a commercial venture. 'Then there are other concepts like regenerative agriculture, where you try to minimise chemicals and plant multi-species crops like longer-rooted pasture species, and you might rest a paddock for a year,' she says.

Sarah accepts that some of her ideas may be bigger than the available budget to implement them, although she firmly believes it is possible to push environmental aspects while maintaining a viable farming enterprise. Unlike Ruth, she seems ambivalent about sheep, but she is committed to coming back to Manangari and working on the farm when the timing is right, and then eventually taking over its management when her mother is ready.

While she is looking forward to the prospect of one day being her own boss, she appreciates running a business can be risky, especially farming. She is also conscious that choosing this life might limit other life choices, such as finding a partner willing to move to the farm, like her father did. However, Sarah doesn't want to be the generation that lets the property slip out of Robinson hands. 'For me, it's always where home is. I know I'm not an Indigenous person, but when I drive up there that's home; it always has been and it always will be. Even though I have moved around the state for different jobs, and have lived in other places, that's *truly* home. My sense of purpose and place and belonging is all connected there,' she says.

Come the first week in November, Sarah will be back at Mannanarie for shearing. She and Catherine are both conscious that Ruth is now in her mid-sixties and that the shearing season takes a physical toll. Sharefarmers still handle the cropping program, and Ruth brings in contractors to do the lamb marking and crutching, but she has generally managed everything else on her own for the past 45 years.

In fact, Ruth's life has never been busier. For the past couple of years she has effectively handled two full-time jobs—managing the farm, and working in the legal practice as many as five days a week because John is short-staffed and rural property sales are booming. With offices in Jamestown, Clare and Port Pirie, and another workspace in Adelaide, John is often away. These days he takes a mate to the opera rather than dragging Ruth along. Prising her away from the farm for any length of time to take a holiday remains a frustrating exercise, particularly during autumn when the sheep have to be fed.

As Catherine says, 'It's her home and she loves it. Unlike some of the rest of us, I think she would be quite happy if she was there with the dogs and the sheep and didn't have to see anybody else much. She loves the farm and wouldn't be anywhere else.'

4

The Dairy Girls

THE CLARK FAMILY, FINLEY, NEW SOUTH WALES

It's rush hour at Glenbank Farm. A large red tractor leaves the yard to fetch a load of silage, followed by a white ute carrying a load of colostrum for some newborn calves. Not far behind zips a four-wheel bike. And then there are the one thousand or so cows, tripping eagerly across the bitumen towards lush pastures after being milked.

This is a fairly typical scene most mornings on the road just outside the Clark family dairy farm at Geraki in the southern Riverina of New South Wales. Sitting on the brow of a low hill watching all the activity is Kristen Clark, who is monitoring oncoming traffic and warning drivers of the bovine hazard ahead. It might look like a quiet country thoroughfare, but within a few minutes a truck rumbles through on its way to a local quarry and three cars pass by, carrying their occupants to work in nearby Finley. Two school buses have already been and gone too.

Sixteen years ago, at this hour, Kristen would have been sitting in Sydney's far more horrendous commuter traffic, making her way to the firm where she worked after graduating from university with qualifications in civil and environmental engineering. That life is far behind her now, set aside to return home and help her mother, Helen, and sister Donna, run the family dairy enterprise.

With three females in charge, the locals have dubbed it Petticoat Junction. The women roll their eyes at this masculine attempt at humour, but it's better than some of the other names applied after Helen decided to face down a broken marriage and one of the worst droughts in Australian history to keep the cows and farm that she loves.

Helen describes life at Glenbank as chaotic. Each day starts quietly enough, in the deep, dark hours of pre-dawn, when farmhands arrive at a quarter to four to bring the herd in for milking and power up the dairy. Before most Australian workers have climbed out of bed, the day has escalated into a never-ending series of routine tasks and spontaneous problem-solving. Tractors, utes, buggies, trailers and four-wheel bikes whiz constantly between the main yard, paddocks, silage pits and the three adjoining properties owned and run by the seemingly unflappable Clark women.

Kristen is usually first on the scene. No matter the season, her day begins around 5.30 when she gets out of bed and makes her way to the modern open-plan kitchen in the house she shares with her partner, Michael, and their two children, Isaac and Xavier. She grinds beans and brews an essential cup of strong coffee in a stove-top espresso maker, then she takes advantage of the peace and quiet to sit at the kitchen table and do some computer work.

It's still dark and drizzling when Kristen leaves the house at six o'clock and walks across the yard to the dairy. A spell of arctic weather is generating news headlines as it sweeps across eastern Australia, bringing freezing winds and rain to the Riverina. Kristen doesn't stay still long enough to feel the cold, but she is wearing a warm high-vis jumper in fluorescent orange and navy, waterproof pants, a padded waterproof jacket and long rubber boots.

The sound of hungry calves, bellowing insistently, follows Kristen across the yard to a large sliding door leading into the corrugated-iron building that houses the milking operation. A flood of light and the smell of fresh manure assault the senses as she steps inside but Kristen is oblivious after a lifetime spent in this environment. In front of her is the raised circular platform that gives rotary dairies their name. Set about a metre off the ground, it holds 50 cows at a time, noses facing inwards towards individual troughs filled with grain. Like a giant lazy susan, the platform turns slowly, presenting the udders of each cow to the shedhands. The design and carefully calibrated speed save people from having to bend or take more than a few steps to attach a milking cup to every teat while keeping a watchful eye for raised tails—a sure sign the cow is about to drop a splattering stream of liquid manure on their heads, or a flood of steaming urine.

Rostered on this morning are three Estonian backpackers, recently arrived in Australia on a working holiday. The Clarks were very pleased to see them. The worldwide COVID-19 pandemic resulted in overseas travel being suspended for almost two years, and Glenbank has been constantly short-handed. With a thousand cows to look after, the farm is a labour-intensive operation that usually requires around ten people to keep up with the daily routine. The Clarks have long struggled to find enough locals willing to take on the work, so traditionally they employ two or three backpackers at a time, to complete the regular team.

Still in their first week, the new arrivals are being trained by Karlo, an experienced employee, originally from the Philippines. Working through the only Estonian with reasonable English, he explains the computerised system that has transformed milking at Glenbank since the Clarks came to the farm almost 50 years ago.

Each animal wears a collar that is something like a Fitbit for cows. Sensors in the device monitor the animal's health, picking up her rumination levels and activity every hour of the day. If a cow's activity level goes up and the amount of time spent chewing her cud goes down, it usually means she is on heat and ready to be mated.

Kristen knows that, on average, a healthy cow in the Glenbank herd spends about nine hours a day chewing her cud. A sudden and severe drop in both activity and rumination is an early indication that she might be ill. In particular, Kristen is on the lookout for mastitis, a potentially fatal disease induced by bacterial infection or sometimes injury, which is the most common health issue in Australian dairy cows. Aside from causing inflammation and extreme discomfort in a cow's udder, it affects milk quality and costs the industry millions in lost production and veterinary costs.

'I don't know how we managed without them,' Kristen says of the collars. 'To tell if a cow was on heat, we used to check visually. We thought we were doing a pretty good job, then we got the collars. They proved way better than we were, and they pick up when cows are just starting to get sick long before we can see symptoms, so we can start treatment much earlier.'

Antennas fixed to a rail running above the milking bails feed the data to a central computer system, which produces daily reports. Kristen reviews them every morning, heading to a messy office space overlooking the platform as soon as she gets into the dairy. Two computer screens sit on a wooden desk scattered with dirty coffee mugs.

A graph on the left-hand screen displays readouts from the collar software. A few clicks of the mouse and a list is produced of animals that are on heat and need to be separated from the herd. An automatic drafting system will change gate settings as each of these cows leave the dairy so they are diverted into a holding yard. Later in the day a technician will come to the farm with straws of semen to artificially inseminate them.

The second screen on the desk displays a list of all the cows currently on the platform, and whether they need treatment of any kind before leaving the dairy. Most critical of all, it identifies when a cow's milk needs to be withheld from the main bulk storage tank either because she is receiving medication or because of some other problem with its quality.

Two more screens showing the same information are placed at head height in the milking parlour where they can be easily viewed by staff responsible for attaching the milking machines. To make doubly sure flagged cows are not missed, a computer-generated voice with an American accent issues a loud warning. To make triply sure, the Clarks have their own colour-coding system which involves spraying symbols across the udders using a non-toxic paint.

A red B warns that the cow's milk has to be captured in a bucket and discarded. A large green F signifies it's a fresh cow, who is being milked for the first time since calving. Her milk will also be collected separately and then fed to the new calves; it is rich in colostrum—a superfood full of nutrition which will give them a better start in life. A blue B means the cow is on a watch list; while she is not being treated her milk will also be captured separately just in case.

Removing the buckets at the end of the rotation is Jo, who has been employed on the farm for almost twenty years. It's a long time to stick at this sort of work, but she loves the cows and 'the girls'. In return, the Clark women appreciate her greatly. Wearing a dark-grey beanie over long brown hair, Jo is wrapped from neck to toe in waterproof clothing, including sleeve protectors and disposable gloves. The last person to see the cows before they back off the platform, she works steadily and calmly, checking that the automatic cup removal system is working and spraying mastitis-affected teats with a fine mist of iodine to help control the infection.

Lists checked and staff given updated instructions, Kristen heads back out into the dark and climbs aboard a farm buggy. She makes for a small paddock near the dairy which holds the springers—cows and heifers close to calving. Calves are born year-round at Glenbank, so the springers are checked twice a day, every day. This morning there are three new arrivals, all female. A small team will return to the paddock in daylight and collect them up in a small enclosed trailer. The calves will be raised by hand until they are old enough to be weaned off milk, while their mothers will be merged immediately into the milking herd.

It's only just starting to grow light when Kristen heads back to the house to pack lunch for the boys and check on Michael, who is not feeling well. It's clear he won't be up to working today, so they confer about allocating his responsibilities to someone else. Michael has been employed on the farm for more years than he and Kristen have been together, and usually looks after feeding out hay and silage. It's a time-consuming but essential daily task.

Back in the yard, Kristen issues fresh instructions to another employee who will be taking charge of the tractor in Michael's place. Then she organises the morning's stock work, which usually involves moving the cows out to pasture after milking, treating sick cows kept behind in the yards, and collecting the newborn calves.

Kristen helps until just before eight o'clock when she rushes back to the house to gather up the boys and escort them to the school bus stop. It's conveniently located next to the farm gate so she hasn't far to go. Her personal list of farm work is endless but Kristen makes time for this moment with the boys every weekday, enjoying their chatter while they wait and quizzing them about what's going to be happening at school. Xavier has been asked by his teacher to bring along some potatoes. He doesn't seem to know why, but his class has been working their way through the alphabet focusing on a different letter each day, and an object to illustrate each one.

Boys safely away, Kristen heads back to the dairy to check everything is in hand as the last of the herd leaves the yards. A problem has emerged with an electric fence so she hops on a four-wheel bike and races off to fix it before going inside for a late breakfast. It's only nine o'clock, and Kristen has already worked more than three hours and ticked off an incredible array of job descriptions—dairy technician, traffic controller, vet nurse, stockwoman, human resources manager, record keeper, fence repairer and mother.

Helen Clark is full of admiration for her daughters. When she thinks about what they have accomplished so far and the scale of the enterprise they are running, she shakes her head in wonder. 'I couldn't manage it now,' she admits.

A city girl, Helen was born in Melbourne in 1949. Her father, Kel Johnson, was a butcher, with his own small shop in the inner-city suburb of Hawthorn, complete with a traditional wooden chopping block and sawdust on the floor. An only child, she lived at the back with her parents, squeezed into four small rooms—a lounge, kitchen, bathroom and just one bedroom. 'That was normal in those days but it was very hard work,' Helen explains. 'We didn't have hot water in the house.

Dad used to boil up water for us to have a bath, and the washing was done in a copper.'

Even while she was at primary school many household chores fell to Helen because her mother, Grace, had chronic heart problems. 'She was in hospital a lot of the time so Dad more or less brought me up. He was a good dad,' Helen reflects.

She had just started her secondary education when Grace died in 1963, at the age of 52. Helen stepped up to take on more responsibility, preparing the evening meal after getting home from school and ironing the white shirts her father wore in the butcher shop.

A social man who loved dancing and a beer after work, Kel went to the pub every night after closing up the shop, staying until six o'clock when drinking laws forced public bars in Victoria to shut. 'Dad was a classic. He'd come home and lift all the lids on the saucepans to see what I was cooking for tea,' Helen recalls. Then after the meal he'd mischievously claim the credit, saying, 'Oh, I've done a good job of that!'

The meals weren't fancy. Meat was always on the menu but it was usually whatever hadn't been sold in the shop and needed to be eaten. To Helen's disgust, that often meant offal such as tripe, brains and sweetbreads—a lovely sounding name for the thymus or pancreas glands from a calf or lamb. 'What a trick!'

By the time her mother died, Helen was attending the prestigious Mac.Robertson Girls' High School near Albert Park Lake. Getting there involved taking two trams—one into the city, and then another out again, travelling along St Kilda Road. The school had a strong focus on academic performance, boasting doctors, lawyers and other leading professionals among its alumni. Helen recalls feeling like 'a square peg in a round hole', but that is not quite how her best friend, Marilyn, remembers it.

'I sat behind her on our first day and we just hit it off. We were the two littlest ones in the form—she was the good one, and I was the naughty one,' jokes Marilyn, who, ironically, went on to become a teacher. 'There were five of us who were really good friends and I never got the sense she didn't belong. She never considered herself a scholar, and she couldn't spell for peanuts, so she got red lines through her work, which probably didn't do a lot for her self-confidence . . . but you only got into Mac.Robertson if you had a certain academic ability and she's a smart lady.'

Part of Helen's problem was that she hated living in the city. She longed for weekends and school holidays when she and her father escaped to Rokeby, a rural community in West Gippsland, about 100 kilometres from home, where Kel bought a 'holiday' house with a small parcel of land. He had grown up on a farm at Korumburra and before moving to Hawthorn ran a delicatessen at Cowes on Phillip Island, so the general region was familiar to him. 'It was only ten acres but we had a couple of horses and a couple of cattle, and it was just lovely.'

Marilyn usually went too, Kel treating her like a second daughter. The two girls spent most of their time simply enjoying each other's company, but as they grew older they each found local boyfriends. 'We didn't give her dad too many headaches. We really were goody goodies. We had the occasional breakouts, but we never did drugs, or smoked, or drank too much,' Marilyn says.

The friendship continued unabated when midway through her Year 11 studies Helen left school. She wanted to move to Gippsland and become a herd tester with the department of agriculture. The Mac.Robertson principal was absolutely horrified. Kel wasn't too happy either, but he eventually gave his permission, provided she found work in Melbourne until she had gained her driver's licence and saved enough money to buy a car. So she secured a job as a photographer's assistant, helping to line up children ready to have their school portraits taken.

At the age of eighteen, Helen became one of two women herd testers in a team of eight based at Warragul, about 15 kilometres south of Rokeby. After a week or so of formal training, she spent a month working alongside the other woman to gain practical experience. 'She had been doing it for a long, long time, maybe twenty or thirty years, when I got there and was probably nearly ready to retire. I was little and she was a very big lady, so it must have looked funny when we rolled up.'

Training over, Helen was assigned a list of 25 farms to visit every month, driving a mini-van that she purchased with her savings. At every farm, she measured the volume of milk produced by each cow and collected samples in test tubes, carefully marked with the relevant cow's number. The tubes were placed in a centrifuge Helen carried with her to separate out the fat content, which was also measured. The results from both a morning and evening milking were then recorded meticulously by hand on large spreadsheets, with Helen drawing on her arithmetic skills to make the calculations.

For stud herds, pleasing overall results could raise the amount farmers received for their animals. Helen soon learnt that this tempted some of the less scrupulous operators to tamper with the samples, injecting extra fat into the test tubes when she wasn't looking. Sometimes they even put two cows' milk into one bucket.

For a young woman working on her own, there was also the risk of sexual harassment. Herd testers usually stayed on farms overnight, so they were on hand for the morning milking. That was mostly okay when the farm was a family concern, but Helen was always cautious if it was operated by a single man. 'There were a few times when I felt nervous and I wasn't comfortable staying. People brushed it off in those days but I thought, "I'll be careful." It might have been perfectly innocent. Maybe they were just being friendly when they put their arm around you and that kind of thing, but I didn't stay there.'

Despite these occasional incidents, Helen loved the work and relished her independence. 'It was challenging and it was also the fact that I was doing it as a woman. It gave me a sense of achievement I suppose. I felt proud of what I did and that I managed it. And I loved being outside—I always have.'

Helen also relished being paid the same amount as her male counterparts—a rarity then. 'I was getting $55 a week, which was a lot of money in those days. Anything over 100 cows was considered a two-day herd so you got two days' pay. I probably had four of those,' she explains.

When she wasn't staying on farms, Helen lived in the house at Rokeby, revelling in having her own space and sharing weekend adventures with Marilyn. Both women joined the Young Farmers club, even though Marilyn was by then attending teachers college in Melbourne and could not take part in many activities.

The club proved a wonderful way for Helen to meet other people her own age. Members gathered regularly for social events, guest speakers on rural topics, farm visits and competitions against other clubs. A club excursion even took Helen and Marilyn to the southern Riverina. 'We drove around all the farms and we thought, "Oh, it's the end of the earth." We couldn't believe the dust,' Helen says. 'We stayed on a dairy farm at Barooga, and we thought, "Fancy putting up with this!", little knowing I'd be living here one day.'

Helen left herd testing in 1970 when she married Ken Clark on her 21st birthday. Ken farmed with his parents at Jindivick, just up the road from Rokeby. The following year Ken and Helen's first child, Donna, was born, and then in 1973 along came Kellie. Within six months the extended family had all moved to the Riverina with 80 or so cows, convinced there were more opportunities in the area because it offered larger holdings and irrigation to sustain pasture year-round.

Ken's parents purchased a soldier settlement block created in the 1950s, on what was previously part of Geraki station. Covering 213 hectares, it offered flat ground, an established channel system to deliver irrigation water and a modern concrete-brick herringbone dairy. There was only one house so the senior Clarks lived in Finley, about 16 kilometres away.

The timing for the move could not have been worse. In 1973 the United Kingdom joined the European Common Market. As a result, Australia's largest export market for dairy products closed its doors in favour of purchasing products from European countries. The following year world prices for butter and skimmed-milk powder collapsed, at a time when farming costs were on the rise in Australia along with wages, fuel prices and interest rates. Inflation rates soared and by the middle of 1974 the country was experiencing a full-blown economic recession. Milk prices and the value of dairy cattle collapsed too. Helen recalls selling a cow to buy Donna a Barbie doll when she was about three and in hospital recovering from tonsillitis. 'The doll was $8 and we got $8 for the cow.'

Recognising the farm could not sustain two households, Ken's parents moved to Melbourne. In 1978 they sold the venture to Ken and Helen, who decided to stay and try to make a go of it. Over the coming years, that meant expanding the number of cows being milked to generate more income.

Apart from a brief time in Melbourne when she tried to keep her father's butcher shop going after he died in 1976, Helen was actively involved in farm life, juggling raising her daughters with milking the cows and looking after the herd. 'It was hard when they were little but I still helped in the shed,' Helen says.

In 1981, the family expanded when Kristen was born. It was a difficult birth, with Helen's general practitioner carrying out a caesarean at the small hospital in Finley. Three years later along came Courtney, after another caesarean. 'Donna was thirteen and Kellie was eleven by then, so I had built-in babysitters. It wasn't planned that way, but it worked out well because the older two always looked after the young ones. While I was milking, they were cooking tea and bathing the kids. It was very handy,' says Helen.

Looking back, Kellie wonders at her parents' faith in their child-caring abilities, given she was so young when Helen left her and Donna in charge of Kristen. 'Would you trust an eight-year-old to bath and feed a baby? I look at eight-year-olds these days and there's not a chance! But that's what she did, and Donna and I used to fight because I loved looking after the babies. Apparently I was also meant to be helping do the vegetables for tea.'

Then Helen and Ken made a decision that completely upended the pattern of their daily lives. They bought a small dairy farm at Lalalty, about 20 kilometres to the east. When Courtney was one, they moved the entire milking herd there and signed up sharefarmers to milk them. Helen's workload might have been significantly lighter, but she felt completely lost. 'It was terrible. I felt like my identity had been taken off me because I wasn't a dairy farmer anymore, and I struggled with that,' she admits.

For the next ten years Helen filled her days as best she could, raising her daughters, volunteering at their schools, visiting elderly people in Tocumwal and planting numerous trees on the farm. She enjoyed having more time to spend with her younger children. What she didn't do was spend more time doing housework, a chore she loathes, or cooking. The latter is a sore point with Helen's daughters, who love to tease her and each other about their different experiences. 'When Kellie and I were little kids, Mum would even make Yorkshire pudding. Kris and Courtney didn't know what a Yorkshire pudding was,' says Donna.

'The big girls used to get desserts, chocolate puddings and all sorts of things, but by the time we came around she was done, which was fair enough,' Kristen says. 'She had a regular repertoire of steak, chicken breast and spaghetti.'

On the flip side, Helen was too busy in the morning to prepare school lunches for her older daughters because she was milking. 'All I wanted was a homemade sandwich. These kids had their lunch made because she had time, and all they wanted was lunch orders,' Donna says.

One by one, the Clark girls finished their secondary education and left home. Courtney recalls that they were given an ultimatum—go to university or work on the farm. Initially, all four chose 'freedom' and tertiary study in the city.

As the oldest, Donna was the first to leave. She went to Melbourne in 1991 to undertake a Bachelor of Applied Science in hospitality at the Royal Melbourne Institute of Technology (RMIT), with vague notions of travelling. As soon as she graduated, Donna found work on Hamilton Island, one of Queensland's most popular tropical holiday destinations. After a year she moved to the Gold Coast, and later on she secured a waitressing job in the heart of Australia at the Uluru resort. She loved the spectacular location and made some good friends among the staff, including the brother and sister-in-law of her future husband, Simon.

Having saved enough money, Donna travelled extensively overseas. When she came back, she visited her Uluru friends, who by then were living in Perth. That's when she met Simon. Initially, they settled in Perth, then Donna fell pregnant with their first child, Eliza, and Simon suggested they relocate to the Riverina for a time. 'You will want to be with your mum and your sisters,' he predicted.

They found a house in Finley, where Simon secured employment at a tyre store. Meanwhile, Donna worked at a bakery cafe in Tocumwal until just before the baby was born in 2001. When Eliza was almost old enough to start school, they returned to Perth so they could, in turn, be closer to Simon's parents, without disrupting her education.

Kellie also headed to Melbourne when she left high school in 1992. Reflecting how much she had loved taking care of her younger siblings when they were babies, she enrolled to study early childhood development. But after twelve months of the two-year course Kellie dropped out. 'I loved the practical side with the children but I hated the theory,' she says.

Returning to Glenbank, Kellie decided to stay and work on the farm. She took on what amounted to apprenticeship training through the Yanco campus of Tocal Agricultural College, picking up useful skills such as fencing, farm bookkeeping and how to artificially inseminate cattle. Over the coming years Kellie gradually assumed responsibility for managing the enterprise, while her parents slipped into semiretirement.

Initially, she lived at the Lalalty property, where the herd was based. That changed in 1995 when the Clarks invested in building a new rotary dairy at Glenbank and brought the cows home. They ended up selling the farm at Lalalty and purchased another property, Orana, just across the road from Glenbank to provide more convenient extra grazing for the herd. Kellie was actively involved in making these decisions, working with her parents on designing the new dairy and revamping Orana, which became her home. They laser-levelled some of the land so it could be irrigated more efficiently, put in a central laneway to make it easier to move the cows around, and constructed new fence lines to create smaller paddocks.

Orana remained Kellie's home after she married Brenden, a Jersey breeder from Queensland, who was sharefarming not far down the road, running cows from his parents well-known Coleshill stud. In 1999, their daughter, Hayleigh, was born. Like Helen before her, Kellie juggled motherhood and running the farm. 'I used to drop Hayleigh at my parents' house at four o'clock in the morning. I'd package her up and I'd take her to Mum's, and when I was breastfeeding she'd bring Hayleigh over and I'd stop the dairy and feed her, then they'd go off again. That's just how it was,' Kellie explains.

As Hayleigh grew older, Kellie took her to childcare every Friday so her daughter had the opportunity to socialise with other children. But apart from when she was milking or drafting cattle, her daughter spent the day out with her on the farm while she worked. The balance shifted a little after her marriage ended; Hayleigh's father lived nearby so she went to his place every second weekend.

The farm enterprise expanded again in 2002 when another farm across the road came on the market. The former sheep property known as Kerry Downs was owned by one of the district's founding families. 'We bought it so Mum and Dad could move off the farm and slow down,' Donna explains.

In 2005, Kellie leased another property just down the road in her own right, with an option to buy. Janeleigh Park was being sharefarmed by her former husband's family, who had decided to leave the district. Kellie ended up acquiring a large number of their Jersey stud cows too, incorporating them into the Glenbank herd.

Meanwhile, Kristen's horizons were expanding. At the age of sixteen, she travelled on her own to South America and spent a year in Guatemala on an exchange organised by AFS Intercultural Programs (formerly the American Field Service).

The global network found her a host family to live with in a town of about forty thousand people and arranged for her to attend the local high school. She had just finished Year 11 studies at Finley, and was keen for adventure before coming back and completing her final year of secondary education. 'It was a dumb idea,' Kristen admits, explaining that while she was away the education department restructured the Higher School Certificate, and she ended up having to repeat a year.

Reflecting on how young she was, Kristen says: 'When I think about it now, I'm not sure I would let my kids do it. I didn't speak any Spanish, and they didn't speak any English, so it was a good way to learn, but I was placed with the family of a local judge, and we just didn't click. Then I became friends with someone at school and went out to visit her family. They lived in a village and her dad worked on a rockmelon plantation. They said I could come and stay with them, so I did. It was a lot nicer, but I couldn't do much at school because I didn't speak Spanish. I attended, but that was just about it.'

After finally finishing high school, Kristen moved to Sydney in 2002. A capable scholar, she gained a Bachelor of Engineering degree, a Bachelor of Arts in international studies and then a Diploma of Engineering Practice at the University of Technology. 'I liked the idea of the practical application of science,' she explains. In particular, she had a vision of returning to the Riverina and using what she had learnt about hydrology and environmental engineering in the agriculture sector. She thought the international studies component might also help open the door to working overseas. Drawn once again to Latin America, she took advantage of an exchange program between universities to spend a year in Mexico as part of her studies.

The youngest of the Clark daughters also couldn't wait to leave the farm. 'I was always told as a kid growing up that maybe farming wasn't for me,' Courtney says, admitting this feedback usually came after she had rushed through a job to get it over and done with, rather than taking care to do it properly, or thinking about the fact that she was being paid and should therefore take it more seriously. She especially detested feeding calves. Unfortunately, she and Kristen were assigned this task every day after school. 'It was the worst job on the farm,' she says firmly, her dislike for the chore obviously still strong. 'I absolutely hated it. They are just so frustrating; they never do what you want and it's very sticky and slimy.'

Courtney escaped farming for a year when she was sixteen and headed overseas on exchange, like Kristen before her. The AFS program took her to Minnesota in the United States. Yearning to travel more after coming back and completing her final two years of high school, Courtney was encouraged by her parents to go to university first. She chose a Bachelor of Business degree focusing on tourism. 'A few years in I realised that I did not need a degree to travel so I completed an advanced diploma in business and worked in a pub for twelve months, saving like you wouldn't believe. I went overseas and worked as a nanny in Spain for six months, then I travelled in Europe for as long as my savings allowed.'

Thirty years after moving to the Riverina, the Clarks appeared to be prospering. Now in their fifties, Helen and Ken were semiretired with Kellie managing an enterprise that had grown from one farm and 80 cows, to three farms and almost 600 cows. Donna was settled in Perth with her own family. Kristen was doing well in her university studies in Sydney, and Courtney was enjoying life in Melbourne. Then in 2005, Helen's world imploded.

She does not like to dwell on the details, but after 35 years of marriage, Helen and Ken separated. 'He decided he didn't want to be here anymore, and he left,' she says succinctly. From her perspective at least, it happened quite abruptly, leaving her in a state of shock. But they managed to negotiate a settlement of their assets, with Helen determined to keep farming. 'Mum pretty much lives and breathes and defines herself by the farm. She'd lost her marriage so she wasn't going to lose the farm as well,' reflects Kristen.

Helen's legal advisers were clearly not sure it was a good idea, given milk prices were low and the Riverina was experiencing what became known as the Millennium Drought. The future of dairying in an area reliant on water from the overstretched River Murray looked bleak. 'The solicitors at first said, "Why would you bother? Why would you do it? Why don't you sell up?" And I said: "I don't want to." And that's how it was,' Helen explains matter-of-factly. 'I had no hesitation taking on running the farm. I wanted to stay. I couldn't have done it on my own but with help I knew I could manage. Kellie was here at the time and we had people working for us so we just went on as was,' Helen adds, underplaying the emotional trauma if not the practical difficulties she faced.

'I think she was just determined to show Ken that she could do it. She was a broken woman for ages but she didn't put aside the work, even though she was broken-hearted,' Marilyn says.

'Mum was a mess. Mentally, she just wasn't herself,' adds Kellie. 'She was devasted because that was the life she had and she knew nothing different, and she'd done the hard yards for all that time. They had [their retirement] all planned, and she didn't see it coming.'

'To be honest I was probably going through the motions for a few years, just surviving, but I wasn't thinking woe is me,' Helen insists. 'It isn't until I look back that I think, how did we get through that? But we managed alright. We still fed everybody, fed our cows, fed the kids. We never did without. We had to cull a lot of cows, but in those days there was drought assistance from the government and we didn't have anywhere near the staff we have now.'

Ken's main interest had always been breeding Holstein Friesian stud cattle and he intended to continue, so they agreed he should take his pick of the cows and most of the heifers. Helen was left with around 420 mainly older animals that would soon need replacing, and a significant amount of debt, mostly relating to buying out Ken's share of the farm.

Word spread quickly in the community about her changed circumstances, but Helen was well respected in her own right. She had been helping to coordinate a local farm discussion group for more than ten years. Every month the group met on a different farm, comparing notes and picking up ideas, with expert speakers providing input on key aspects of managing a dairy business. 'I had learnt so much through that, just

Amber Driver with her husband John and their children Ruben (far left) and Sonny.

Looking across the beautiful Elkedra River to the homestead (image by Liz Harfull).

LEFT: Joan Driver with Roy and Dennis, 1950 (John K. Ewers collection, State Library of Western Australia 274519PD).

Amber with Scotty the cowboy.

ABOVE: Jerry Killen walking Amber down the 'aisle'.

RIGHT: Roy and Catherine Driver.

ABOVE: John and Sonny Driver.

LEFT: Amber flying the station Cessna (image by Liz Harfull).

BELOW: Part of the luscious green lawn surrounding Elkedra homestead.

Kelly Dowling counting sheep in the Coolong woolshed (image by Liz Harfull).

Storms over Coolong.

LEFT: Kelly with her brothers, Robbie and Luke.

BELOW: Kelly (far right) with her Kinross mates—(from left) Tara, Kathleen and Lynn.

Kelly shaking the hand of Prime Minister John Howard during his tour of drought-affected areas in 2006, her father Eric and son Ned looking on (image by John Feder, Newspix).

ABOVE: The Dowling family on Christmas Day—(from left) Theresa, Luke, Sam, Laura, Kelly, Eric, Phil, Ned and Kim, with Penny in front holding Robbie, next to Poppy and Barney the dog.

LEFT: Kelly and Phil with their son, Ned.

Nancy Robinson.

TOP: Ruth Robinson with her sheep at Jamestown market, December 2022 (image by Kate Jackson, Stock Journal/ACM).
ABOVE LEFT: Ruth on her first day of school, with her older sister, Mary.

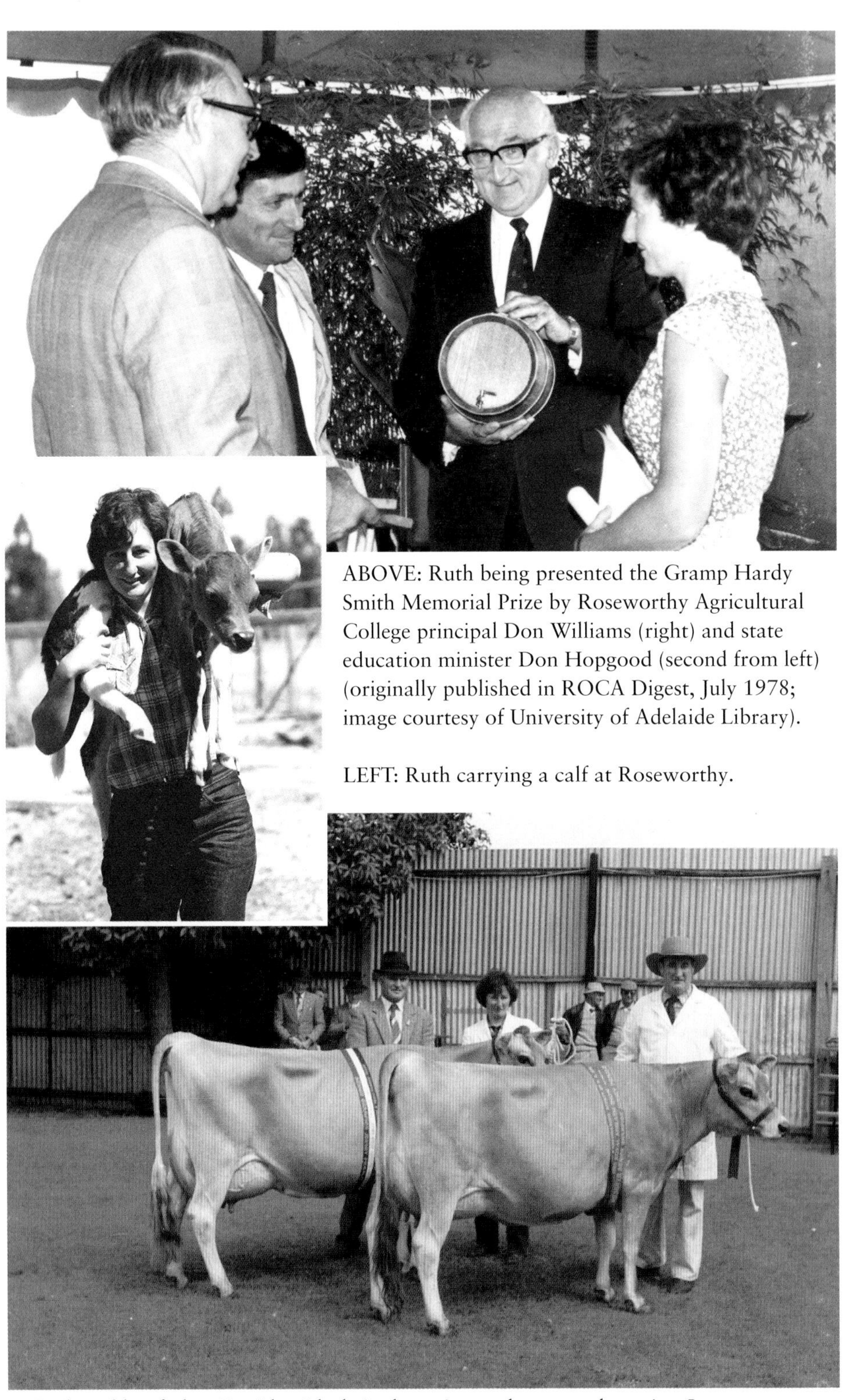

ABOVE: Ruth being presented the Gramp Hardy Smith Memorial Prize by Roseworthy Agricultural College principal Don Williams (right) and state education minister Don Hopgood (second from left) (originally published in ROCA Digest, July 1978; image courtesy of University of Adelaide Library).

LEFT: Ruth carrying a calf at Roseworthy.

Ruth and her father, David, with their champion and reserve champion Jersey cows at the Royal Adelaide Show.

RIGHT: Ruth and John on their wedding day, April 1985.

BELOW: Ruth's daughters, Catherine (left) and Sarah.

Helen Clark with her daughters Donna (left) and Kristen (image by Liz Harfull).

ABOVE: Helen's parents, Kel and Grace Johnson.

RIGHT: Helen (left) with her best friend, Marilyn.

BELOW: Kristen leading a calf.

RIGHT: The Clark sisters ready for school—(from left) Donna, Courtney, Kellie and Kristen.

BELOW: Three generations of Clarks—(from left) Donna holding Abbey, Eliza, Kristen, Helen, Courtney, Hayleigh and Kellie.

Helen sweeping out the barn with help from Eliza.

LEFT: Belinda Williams and Michelle O'Regan.

RIGHT: Belinda with her mum, Pam (image by Liz Harfull).

LEFT: Carolyn 'Caz' Kingston-Lee and helper picking tomatoes during a visit organised by Whitsunday Regional Libraries in partnership with the PCYC (image courtesy of Stackelroth Farms).

ABOVE: An aerial view of the farm just before Cyclone Debbie hit.

RIGHT: Michelle's favourite image of Belinda's workworn hand.

BELOW: The picking crew at the end of the 2019 Halloween pumpkin season (image courtesy of Stackelroth Farms).

Instagram moment: Michelle posing Belinda with sunflowers to promote their Mother's Day event (image by Liz Harfull).

Nancy Withers with Pomanda Titan (image by Ann James).

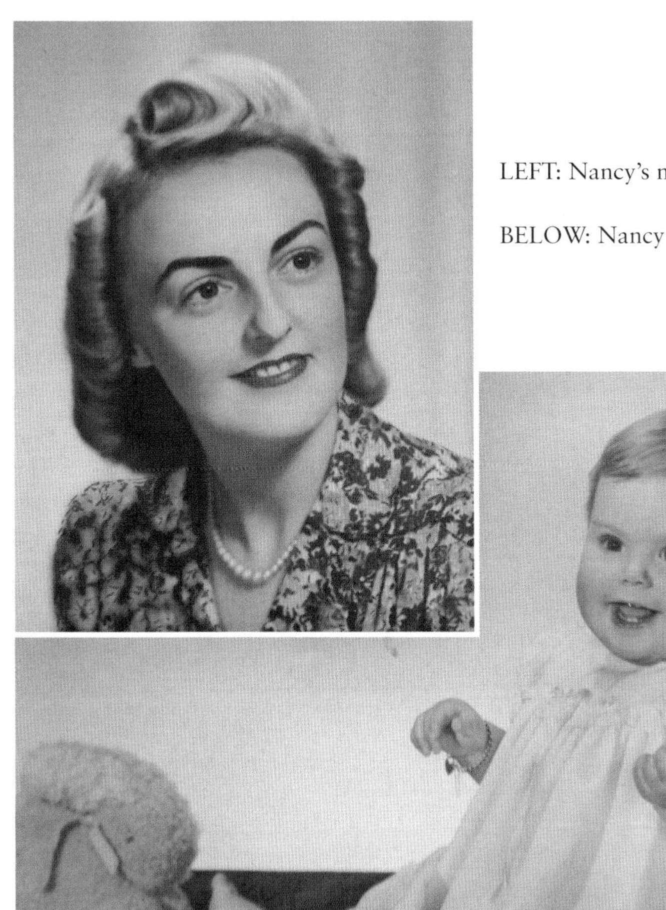

LEFT: Nancy's mum, Bette Mitchell.

BELOW: Nancy as a baby.

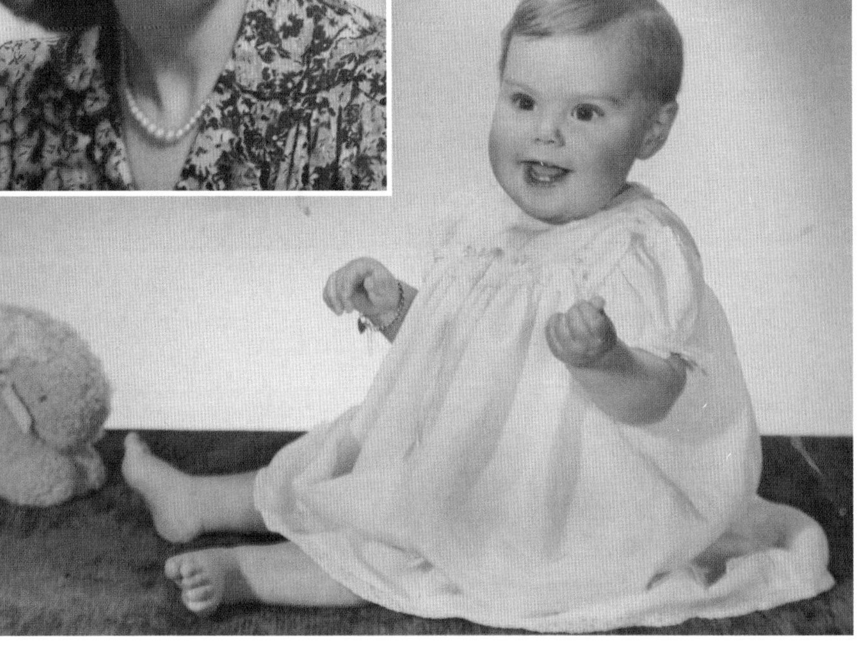

RIGHT: Nancy riding Gypsy on Nalpa, with her dog, Shandy.

LEFT: Nancy and Bullenbong Mate.

BELOW: Nancy working in the yards at Tupra station with Pomanda Iago (image by Margie McLelland).

listening to other farmers and seeing what they were doing. That helped me a lot and it gave me confidence to make decisions . . . and I could pick up the phone to one of them if I had to.'

While Helen knew a great deal about managing the herd and the farm finances, there were other tasks she had not been as intimately involved in, such as selecting pasture varieties and choosing bulls for the AI program. The most immediate challenge was feeding the cows at a time when the paddocks were bare because of the drought. To realise her full potential when it comes to milk production, a single Holstein Friesian dairy cow on Glenbank chomps her way through almost two tonnes of green feed, more than 5 tonnes of silage or hay, and another 2.8 tonnes of concentrated feed such as grain every year.

During the Millennium Drought, water allocations from the River Murray for irrigation sometimes dropped to zero and there was very little rain to sustain pastures. The year 2006 was one of the driest on record for many parts of southern Australia, devastating communities and the environment. Desperate for feed that would keep her cows milking, Helen was forced to buy in hay and silage, which was expensive and not always the best quality. She still shakes her head over one of her biggest mistakes—paying $80,000 for a load of faba bean silage. It had been tested for energy content and the results were good so Helen confirmed the order. When they opened it up it was rotten.

Another repercussion of the drought was Kellie stepping down as manager. With mortgage repayments to make on Janeleigh Park and things tight on Glenbank, she signed up to work, fly-in, fly-out, for a coal-mining company in Queensland.

For around two years, this involved her spending two weeks away at a time, with a week home in between. Employed as a field officer, she coordinated accommodation for exploration crews, as well as working on mine sites. Although there were a few female geologists, it was very much a male-dominated environment which didn't faze Kellie, and it marked a turning point in her life.

'That was basically my breakaway from the farm. It's when I realised there are easier ways to make a living,' she says. 'Don't get me wrong, I loved my time on the farm and I had a good life, but you could knock off at five o'clock [in the mining job] and there was no stress. It was a simple life.'

By 2012 Kellie had put farming behind her and moved to Darwin. She sold Janeleigh Park to her family and purchased a hectare near the tropical city instead. A few years later, she took over a steel fabrication engineering business, employing ten boilermakers. Her right-hand man was her partner, Noel, whom she had met years before in Queensland. While it was still a male-dominated environment and she once again had a business to manage, Kellie still reckons it was a simpler life.

When Kellie took on the mining work, it was Donna who stepped up to replace her on the farm with support from an employee, Pat, who had been helping with day-to-day management of the enterprise for some time. Donna and her family had moved back to the Riverina soon after Ken left, so they were on hand to support Helen. Struggling with her own emotions around the breakup, Donna was shocked to see how much weight her already diminutive mother had lost.

'I suppose we are very controlling women, very controlling, and it was something she couldn't control. It was something

Mum couldn't fix,' she adds frankly. 'Even though I was 33 at the time, I didn't understand what was going on. You think your parents don't have feelings and emotions—they are strong, they have no needs—and I probably didn't handle it very well either.'

Initially, Donna found casual employment back at the same bakery cafe where she had worked before Eliza was born, and helped on the farm when needed. After choosing a career in hospitality, she had never really planned to come back to the farm permanently, but when her mother asked in late 2006 if she'd like to take on a managerial role, Donna agreed. 'It wasn't a hard decision at all. I was only working part-time at the bakery and this seemed the best decision for my family.'

Courtney returned to the farm around this time too, working alongside Donna and her mother for about eighteen months. After coming back to Australia, she had taken up her old pub job, but the place had changed and it was costing too much to live in Melbourne. 'So with Kellie leaving, I decided to come back and help while I worked out what I wanted to do.'

Courtney took responsibility for the bookwork and paying accounts, and helped Donna with the milking. 'It was a steep learning curve. I was only 22 and I didn't know how to run a farm. Being the youngest, I just did what I was told,' she says. It didn't take her long to realise that being involved wasn't a long-term option. 'I decided farming wasn't really my thing, and I like having my mum as my mum and my sisters as my sisters, not as my bosses!'

Eventually Courtney joined the public service, taking up postings in regional Victoria after another brief spell in

Melbourne made her realise that she was over city life. 'After not being able to wait to leave as a teenager, going back as an adult I couldn't wait to get back to the country. As soon as I could get out of the city I did,' she says.

Kristen came to the same conclusion. After graduating in 2007, she was offered a full-time job in Sydney with an engineering consulting firm, working in the environmental sector on hydrologic modelling, and monitoring river systems and flooding patterns. 'It was alright. I just didn't like sitting in an air-conditioned office with no windows at a computer all day. That's when I decided it wasn't all it was cracked up to be. You commute for 40 minutes a day to sit inside an office and everything is a hassle—finding a house to live in, living in a pokey little house, all the big things.'

Kristen returned in 2008. For two or three days a week she worked for a regional water authority based at Tatura. The drive took 90 minutes so she stayed overnight to reduce the travelling, and spent the rest of the week helping Donna manage the farm. Then she took additional work as an independent consultant with a government-funded water project. 'It was good money but it was a dumb idea,' she says. 'It all got too much, doing three jobs—surprise, surprise!—so I gave up Tatura.'

Within a year or two, Kristen pulled the pin on the consulting work too. 'I just got burnt out I guess,' she says. 'It was a big project and I was out of my depth, and it just wasn't much fun. It would be a lovely day outside and I was sitting in the house at a computer. It just sucked. I wanted to get more involved in managing the farm and actually doing things. It was a well-paid job, but there was money to be made on the farm too.'

Along the way, she and Michael became partners. They had gone to school together in Finley and Kristen was a friend of his sister's, but he was a couple of years older so she didn't have that much to do with him. Life and other relationships ebbed and flowed, Michael came to work on the farm and they both found themselves single.

After ten years together, they are adept at juggling busy routines. Still a full-time employee, Michael happily takes responsibility for many household chores as well as farm maintenance and most of the tractor work, leaving Helen, Donna and Kristen to make the management decisions. 'He's our token male on the farm. We shouldn't be sexist but we tell him he has to do all the boy jobs,' Kristen jokes, later adding: 'We do rely on Michael to keep everything going. When the wheels fall off, he's on call to get us out of trouble. If we separated, Mum and Donna would probably keep him and send me packing!'

The Clark women insist none of their partners have trouble recognising their authority when it comes to running the enterprise. 'If I had stayed in my job as an engineer, my partner wouldn't think that they had a right to walk into the office and have a say in things. It can be messy with family farms, but we make the management decisions,' Kristen says. 'Michael has more autonomy [than other staff] and we do consult him on some things but we didn't ask him, "Do you think we should build a shed?" We said, "We are building a shed." And he's on board,' she adds.

Helen found a new partner about fourteen years ago, who also keeps out of management decisions. Rodney is a beef producer with farms of his own, and a property at Cobar that he runs with his brother, so he is constantly on the road. 'He is a very tolerant, patient man. He's very calm, unlike me!' Helen says. 'Our farming lives are very separate. I don't have anything to do with his, and he doesn't have anything to do with mine, and it works well. It would have been hard for the girls to have someone come in and want to change things, but that was never the case. Rodney knows not to interfere.'

Simon is similarly aware. 'He did work here for a short time and it did not work. It ended in tears,' jokes Donna. Not from a farming background, Simon runs his own tyre-service business in Finley. He was managing the store of a well-known national tyre company, but shortly after the pandemic started it temporarily closed all regional stores, even though they offered an essential service. The company eventually decided to close the Finley business permanently and wanted Simon to relocate but after years of commuting he wasn't interested in managing tyre stores in Victoria. 'It's been hard but it's going okay, and he's done it all by himself,' Donna says.

All three women agree most of the men encountered in their working lives have also been respectful of their authority, but it isn't always the case. The most common offenders are cold-call salesmen. 'They will drive in and if there is a man standing next to you, they will look straight at the man and talk to him. They assume he is the farmer. They did that to me a bit, but Kristen has noticed it too. In fact, I think she has found it more [of an issue] than I did,' says Helen, who thinks part of the problem is her small stature. 'Many years ago, somebody came to the door and asked me was my father here. I said, "I hope not." He'd been dead for years!'

It's eight o'clock in the morning and Helen is in the hayshed on Kerry Downs, where the youngest calves are kept on soft, clean straw in individual pens. Helen is responsible for raising around 450 calves every year, a workload that has increased with staff shortages, despite her daughters' encouragement to slow down.

For some years Helen was only responsible for feeding them a few days a week, which enabled her to look after Kristen and Michael's boys, Isaac and Xavier, during the day, until they reached school age. Then the experienced farmhand who had taken over from her found a job in aged care. At the age of 71, Helen returned to working full-time for almost twelve months, spending at least six hours almost every day tending calves, with support from Donna.

Helen is reluctant to step back completely from supervising this important aspect of the enterprise because it takes skill and experience to make sure the calves thrive. Most days of the year there are around 150 to care for, starting with her 'babies' who are fed individually morning and afternoon, slurping from a rubber teat attached to a large plastic bottle. It's a messy business, with the calves slobbering over Helen as well as the teat. Getting them to settle and start drinking is surprisingly difficult, so she usually climbs into their pens and holds them in place between her legs. This can lead to quite a tussle, especially with the large Holstein calves who come close to outweighing her within a few days of birth and sometimes knock her over in their enthusiasm.

Donna helps with the calves in the morning and one of the farmhands delivers milk from the dairy in a 300-litre green plastic tank set on a small trailer. It's emptied into large tubs so a special powdered supplement can be whipped in, using a modified battery-powered concrete mixing tool that Michael put together.

After seven days, the calves are moved into small shared pens alongside the hayshed, where they learn to drink from a trough. Then when they are two or three weeks old they are aggregated into larger yards with calf shelters and access to grass as well as milk. They stay there until they are three months old, when they're weaned, tagged, wormed and vaccinated. Helen, Kristen and Donna usually tackle these tasks together once a month, working in the stockyards with a well-established rhythm accompanied by lots of friendly banter.

An old plastic outdoor table stands in one of the yards and on it sit two types of eartags: small brass tags engraved with unique numbers that are kept on file in the farm records so the Clarks can trace the details of every animal throughout their lifetimes on the farm, and official National Livestock Identification System tags that must be attached to every animal's right ear to meet government regulations. There is also paraphernalia to dose the calves with vitamins, worm drench and a seven-in-one vaccine that will immunise them against a range of potentially fatal diseases such as pulpy kidney, tetanus and leptospirosis.

Helen is in charge of noting down the numbers and making sure every heifer is accounted for, while Kristen attaches the tags and administers doses. Donna is responsible for pushing the animals up into a narrow raceway fitted with a cattle crush. Encouraging them to move along, she waves her arms and calls out loud 'hup, hups'. Her voice is husky—something she blames on yelling at cows and calling out across open spaces. 'That's the Clarkacom! Donna's husband jokes that we don't need CB radios, we just screech from afar at one another,' Kristen explains with a grin.

Watching them work together, it is clear Donna and Kristen have an easy camaraderie. There is frequent joshing while they confer on tasks, whether it's via their mobiles or crossing paths while moving from one job to the next. It took a little time for them to fine-tune their working relationship after Kristen came home. She was only eight when Donna left, so they had not really spent all that much time together growing up. 'It probably took us a couple of years to really settle in, and we were learning a lot about running the farm at the same time,' Kristen admits.

Part of the challenge was that the sisters have very different ways of working. 'Donna wings it while I like to pre-plan everything. Now it's easier because we know what we are like. We have probably had our differences over time but we have learnt to compromise on our own way of wanting to do things, and accept the other person's way. In hindsight Mum was very patient to put up with us, but we always say communication is probably our biggest strength.

'We have a meeting once a week with the three of us and we sit down and talk things out. Even when we don't agree and it gets tense, we always sort it out. We don't let things fester. If there is a problem, we sort it out at the time rather than holding onto resentments for months. And we have settled into roles where our strengths lie.'

Helen agrees. 'I can't believe how well these girls work together . . . It's a hard job to do and you have to really get on well, and these girls are not similar . . . It's worked way, way better than I would have thought.'

Donna says it's the best possible combination of sisters even though she and Kristen have very different personalities. Kristen describes herself as more of an introvert, and the family tease her about how much she likes spreadsheets and putting together lists. She is quite happy looking at data and analysing how the business is performing, and every year she creates a fodder budget, estimating how much feed the herd will need and the most cost-effective options. 'I would happily never talk to anybody and just do my job, but that's not how the world works,' she says.

Donna, on the other hand, is considered more of a 'PR' person, better at dealing with people and comfortable making small talk. She's pretty sure she annoys Kristen when they sit down for the weekly meeting because she likes to 'gossip' when they are meant to be focusing on business. For Donna, having a sense of humour is the most important thing when it comes to recruiting staff. 'I always say to Kristen, "You can employ that person if they are going to make me laugh".'

Managing staff once they are on the payroll is a big part of Donna's day, along with wrangling children. Her routine begins at around six o'clock when she gets up to enjoy a quiet breakfast on her own, does the dishes and puts on a load of washing. After a gap of eight years, she and Simon had another daughter, Abbey, who was born in 2009, and a son, Hamish, who is eleven. Abbey is up by seven, but it takes longer to rouse Hamish out of bed and into the shower. 'The hardest part of my day is getting that boy to school,' Donna jokes.

Most mornings, she juggles this task with feeding calves, coming home to make sure both children get on the bus that collects them from in front of their house. The rest of her time before lunch is usually taken up with stock work, or stepping in to help with anything that needs to be done. On Mondays she pays farm bills, and on alternate Thursdays she pays the weekly wages, taking turnabout with Kristen.

After lunch, Donna heads to the dairy to help set up for the afternoon milking. Then at three o'clock she takes over supervising the rest of the day's work so her sister can finish up. They worked out long ago that it made sense for them to stagger their days so there is always someone on hand to deal with any issues that arise. This also enables them to work a more or less eight-hour day and take alternate weekends off, a rarity in farming.

Perhaps Donna's most frustrating chore is preparing the staff roster—a seemingly endless process exacerbated by COVID-19 and backpackers resigning at short notice. On a cold June morning in 2022, it's happened again. After a week on the job, the Estonians have decided to leave. The leader of the group approached Kristen after the morning milking to explain they have been offered more money per hour somewhere else. They have decided to take it because they are trying to save enough to return home for a family wedding.

Kristen is extremely disappointed. 'You get these people and spend time and money training them, and then they go . . . It's so frustrating,' she says. 'We got an Argentinian guy, and he was here for a couple of months and then he said he had an opportunity that was too good to miss. Then we had an Irish girl who was here for ten days, then she left for the Northern Territory. It makes it sound like this is a dreadful workplace. Milking cows is not the best job in the world but we don't screw people over. We pay above award wages and we try to do the right thing.'

Later on, Kristen discovered that the root cause was most likely an Australian government scheme that operated during the pandemic years, giving backpackers a payment of $2000 to cover the costs of travel and accommodation when they left home to take up short-term agricultural work. To claim the money, they had to complete at least 120 hours work over no fewer than four weeks. 'It turns out some of them were coming with no intention to stay. They worked 120 hours to qualify, and then left!'

Kristen reflects on earlier employees from seemingly unlikely backgrounds who loved their time on the farm, and have kept in touch. 'We had one girl who sold perfume in France and you would have thought she would be the last person to settle in to working in a dairy, but she was great and she loved it. And a German girl who was a physio liked it so much she decided she didn't want to go back to being a physio and was going to find work on a farm instead.'

Back in the house, Kristen puts another pot of coffee on to brew and makes some porridge. Then she opens up her laptop and immediately starts looking for replacement labour. It might take days to find someone and the Estonians have given limited notice, so there is no time to waste.

The best place to start is Facebook where several pages are dedicated to backpackers looking for work in Australia. One site alone has more than 200,000 followers, including potential employers. The trouble is most of the travellers are looking for work in the warmer climes of Queensland or the Northern Territory. They dream of adventure in the wild outback or serving drinks in a bar on the Gold Coast, not dodging showers of manure in the cold and dark. To make matters worse, changes to Australia's working holiday-maker visa program in recent weeks mean they no longer have to work a certain number of days in agriculture to meet the terms of the visa. Farm businesses are now competing against the tourism and hospitality sector.

So, the Clarks are going through the process of organising two more workers from the Philippines, under a federal government employer sponsorship scheme. They are friends of Karlo's, with experience working on dairy farms. The application process takes time, and costs thousands of dollars in fees, but Kristen reckons it will be worth it because they tend to stay longer, with a recognised pathway to becoming permanent residents.

Even if the Filipino workers pan out and the staffing pressures ease, the Clarks plan to consolidate the farming operation in the immediate future, increasing the herd by only a relatively small number compared with previous growth, and improving infrastructure. The enterprise has doubled in size since Helen took over in 2005, despite setbacks during the Millennium Drought and another severe drought about four years ago when their water allocation for irrigation was cut to zero and they had to cull cows. By 2022, the Clarks were milking 950 cows and producing almost 9 million litres of milk a year—that's enough to meet the average requirements of more than ninety thousand Australians, just from one farm. The business targets they have set themselves allow for the milking herd to increase by another hundred, and they are constructing a massive new open-sided shed where sick cows can recover and springers can calve under shelter. Kristen has wanted to do it for years, but the others were cautious because of the cost and effort involved.

The family is very conscious that while they have hung in there, the number of dairy farms in the area has more than halved since Helen took over. It makes the girls admire their mother's achievements even more, and her willingness to give them a go.

Part of Helen longs for the days when she was milking just 80 cows, and they all had names. She is worried that her girls work too hard, while admiring all that they have accomplished. 'I'm very proud of what they've done. They have proved their worth. There are a lot of things I would probably have boo-hooed or not done, but they have had the confidence to go ahead. I'm fairly conservative and I suppose I am a pessimist. I always see the worst that could possibly happen, but youth are brave,' she says.

'I don't know,' Donna remarks while she is feeding calves. 'I couldn't do it without Kristen or Mum. I couldn't do it by myself, it would be way too stressful.' But she has absolutely no regrets about coming back to the farm. 'This is my life. I don't like feeding calves, I don't like milking cows, I don't like sorting cows, I don't like chasing cows but I love it all. I don't want to do anything else. I never ever wake up in the morning and think, "Oh no, I've got to go to work." I couldn't imagine a different life. And we are so lucky to be working with Mum, and to be with her every day.'

Adds Kristen: 'We don't reflect on it a whole lot but it is

pretty special, the three of us here together. And we are very lucky that Mum gave us the opportunity.'

Speaking on the phone from Darwin, where she and Noel are living once again after twelve months travelling western Queensland doing construction work, Kellie is worried that her mum and sisters are working too hard. 'They are my heroes. I admire each and every one of them, and probably Mum most of all, but I keep saying to her, "Mum, you've got to stop, you've got to enjoy life while you can." She has achieved a lot in her life. I suppose you could say she has built an empire. She doesn't see it that way, but I do. I think it's amazing what she's achieved under the circumstances. Mum is the rock for all of us. She is the rock.'

In 2015, the family went through a succession-planning exercise, to work out where they were heading, and to make provision for Helen's retirement. They agreed that by 2025, she would step out of managing the business. Not that Helen is planning to stop working entirely. 'In a few years' time I want to be doing something, but not as much as I do now. You have to have something to get out of bed for in the morning, just not seven days a week,' she says.

Kristen can see the day when all three of them step back, and hire managers to run the enterprise. 'You wouldn't just get anybody but you could conceivably have someone manage the enterprise, and not just sell up if the kids choose not to come back,' says Kristen.

She and Donna are both adamant that they don't want any of their children to feel obliged to stay on the farm and run it. 'I'd love them to, but you can't put that pressure on them,' Donna says.

'Parents' expectations shouldn't be the reason they come back,' agrees Helen, anxious about the idea that any of her daughters might have felt that way. 'I hope they didn't come home because they felt they had to. I don't want that on my shoulders,' she adds sternly.

That's not to say Helen isn't excited about the prospect that at least one of her grandchildren is open to the idea. At the age of twenty, Donna's eldest daughter, Eliza, is passionate about the future of Australian agriculture. In 2021 she was awarded a prestigious Dairy Australia scholarship to study for a Bachelor of Business at Marcus Oldham College in Victoria. The Geelong-based college is one of Australia's best tertiary institutions for the agriculture sector and her three-year study program focuses on managing a dairying business.

Eliza made the decision to enrol after spending her gap year working full-time on the farm. She even took part in the Monday-morning meetings so she could learn more about the management side. The experience gave her a different perspective from just helping out on weekends or after school. Uncertain what career path she wanted to follow when she left high school, she was soon convinced that her future lay in agriculture.

When she told Kristen, her aunty suggested Marcus Oldham and encouraged her to apply for the scholarship. Eliza didn't expect to be successful, but she looks up to Kristen as a role model and took her advice. 'Really, I thought "I'm never going to get this", but I have a lot of women in my family who have achieved so much and I'm so proud of them, that it gave me a lot of confidence,' Eliza says.

That confidence has spilled over into her studies, where Eliza is one of only ten young women in a class of 40, and thriving: 'It's the best decision I've ever made because I absolutely love it,' she says. In particular, it helped her to find her feet when she went to work for one of Australia's largest dairying operations during a year of practical placement. The Glenbank enterprise is large but Moxey Farms, near Forbes, employs around 220 people and milks thousands of cows in a state-of-the-art set-up with a reputation for innovation.

Growing up, Eliza never really considered there was anything surprising or unusual about women running a farm because she saw it every day. But since leaving the district where the Clarks are well known, she has regularly found herself correcting people's assumption that she is following in the footsteps of her father. 'They are always so surprised that it's all women, and when I tell them that it's a large herd as well, people are normally shocked, and then they have a lot of questions,' she says.

Eliza is particularly proud of her grandmother for keeping the farm going after her husband left, and the way she has worked with her daughters to build such a successful enterprise. She admires her Aunty Kristen for her business brain and desire to learn new things, and she is extremely impressed by Donna's 'superwoman' performance raising a family, running the farm and still finding time to cook dinner every night. 'She never complains. She just gets up every day and gets on with it . . . She says she's the gossip and the disorganised one, and that's what she thinks, but Mum is the one who keeps everything calm and sane. She keeps it under control,' Eliza says, then adds: 'They all have different attributes that they bring to the farm, and it all just ends up working together very nicely.'

Although she knows Helen, Donna and Kristen are thrilled that she wants to work in agriculture, Eliza doesn't feel under any pressure to return to Glenbank. 'Even if I don't, I know they will still be proud of me and what I do. And if I don't go back, I will always want to be part of the farm because I want to see it go on, and I always want it to be girls! That's maybe why I want to go back, just to carry on the tradition,' she says.

5

Women Can't Farm!

BELINDA WILLIAMS AND MICHELLE O'REGAN,
BOWEN, QUEENSLAND

It was deceptively quiet inside Pam Stackelroth's house when Cyclone Debbie hit Bowen in 2017. Outside the monster tropical weather system roared like a train, wreaking havoc. Exhausted after days of preparing for the onslaught, Pam's daughter Belinda and Belinda's partner Michelle tried to snatch some sleep, confident the house would withstand the ferocious winds, but fearful of what might be happening out on the farm. There was a lot at stake. Just five days before they had finished planting 30,000 butternut pumpkin seedlings.

Moving incredibly slowly, the tropical cyclone stayed centred over Queensland's Whitsunday region for much longer than normal, generating gusts of up to 260 kilometres an hour and surging seas that tore apart towns and farms; but worse was to come. The next day a massive electrical storm brought torrential rains. The already saturated ground became a quagmire. Belinda knew it would start turning to concrete as soon as the baking tropical sun emerged, making it extremely difficult to pull out hundreds of metres of damaged plastic and trickle tape. To avoid a nightmare situation, they had to act quickly.

What happened next reflects the esteem in which Belinda and Michelle are held, and the generosity they have shown others in their community. It's a recurring theme in the lives of the two women, who have found a unique way to combine their passions in life—Belinda, a third-generation farmer who loves growing vegetables, and Michelle, a police officer who works with disadvantaged youth.

Belinda Williams and Michelle O'Regan were born within a month of each other in 1973 and went to the same high school, but those who don't know better might say that's where the similarity ends. Introverted and self-effacing, Belinda was raised in a farming area set on a rich alluvial plain only a few kilometres out of Bowen, where she has been a farmer most of her life. Lean and quietly spoken with a wry sense of humour that's easy to miss, she is far happier out on her property tending to the vegetables she grows for a living than facing a group of strangers.

Michelle is a self-confessed extrovert, who loves meeting new people and often takes centrestage at local events. Generous in body and spirit, and quick to show emotion, she sees potential in even the most recalcitrant teenager and willingly shares her personal story if she thinks it will inspire better life choices. Together for almost twenty years, these two women share not only their love for each other, but childhoods where the kindness and concern of others made a huge impact.

As she explains it, Belinda Williams, known to her friends as Berl, was born on the hill in Bowen (the local way of referring to the town's hospital), and got as far as Delta—a small community about eight kilometres west of the town, which was at one time mostly small farming blocks where families grew fruit and vegetables for a living. Her parents, Graham and Pamela, grew tomatoes in partnership with Graham's parents, who lived in a little house at the other end of the property. Somewhere in the family collection is a photo of Belinda in a nappy playing in a muddy irrigation channel while Pam is planting out seedlings. 'So the soil is in her soul,' says Michelle.

An only child, Belinda didn't like to sleep much during the first two years of her life, keeping her mother well and truly on her toes as a very active toddler. 'Mum thought the first day I went to school and they asked me, "What's your name?" I'd say, "Jesus Christ", because that's what my mother was always saying. "Oh, Jesus Christ, what are you doing now!" She couldn't even park me in front of the TV; I always had to be doing something.'

One day that something was pedalling her tricycle into town. It had squeaky wheels, but Pam had refused to let

Belinda's grandfather oil them because the sound was a useful indicator of her whereabouts. On this particular day the system failed. When Grandad left for town, Belinda followed. She was found at the end of the farm's long driveway, near the bridge that crosses the Don River.

As she grew older, Belinda began helping on the farm. Before catching the bus to school in Bowen, she often worked in the shed, folding cardboard packing cases and securing the bottoms using a big foot-operated stapler. She helped out after school too, and at weekends. Then, when she was fifteen, her parents sold the farm and moved to Mooloolaba, on the Sunshine Coast, where they planned to buy a business. Belinda stayed put in Bowen to finish her Year 10 studies before joining them, living in town with Pam's parents during term time, and then travelling to Mooloolaba for school holidays.

Before Graham and Pam found anything suitable, a newsagency in Bowen came on the market so they bought that instead and moved back. Helping out in the business became a part-time job for Belinda while she was still at school, and then more or less a full-time job after graduation. It wasn't her first preference, but home life was becoming increasingly fraught. 'My old man was a bit of a heavy drinker, so Mum and I were protecting each other,' she says, shying away from revealing any more detail.

When Belinda was about eighteen, her parents separated and sold the newsagency. The new owners needed someone to show them the ropes, so she agreed to stay on while also helping out every now and then on a farm operated by Ian Stackelroth, who became Pam's new partner. A former police officer known as Stacky, he was a highly respected fruit and vegetable producer with land close to where Belinda grew up. Within a few years, she was working on the farm full-time, assigned to one of the picking crews. She loved the experience and working with her mother and Stacky, whom she admired greatly.

Aside from their own growing operation, which employed about twenty people during peak season, Pam and Ian were part of Climate Capital Packers, a substantial horticultural enterprise operated in partnership with two other growers, Peter Reibel and Peter Collyer. By combining forces they had more resources to develop new markets, and greater leverage when selling their produce and negotiating prices for essential inputs. The company also operated its own commercial nursery, a processing plant, and a packing shed that employed 30 to 40 people and distributed produce nationally.

To expand their own production, Ian and Pam leased the property which is now the home base for Stackelroth Farms. They also leased another block about 30 kilometres southwest of the home farm, on the road to Collinsville. This gave them a total production area of around 97 hectares, which is substantial for the horticulture sector, with space for large crops of sweet corn, butternut pumpkins, watermelon, rockmelons and capsicums.

The season was in full swing when tragedy struck on a cool Wednesday in September 1996. Ian, Pam and Belinda had been up at the far block with a team of pickers. The two women left ahead of Stacky to drive the crew back to the home farm at the end of their shift. They became concerned when dusk approached and Ian had failed to return so they went to check.

Belinda found him first. Ian appeared to have tripped or fallen while carrying a loaded firearm. Belinda believes he was most likely going after some feral pigs that had been damaging crops in the area. Whatever happened, the gun had gone off, injuring him fatally.

Ian was only 43 when he died, leaving a gaping hole in the lives of Pam, Belinda and the farm. In many ways the next few years are a blur to Pam, but the first weeks were taken up with the urgent business of finishing the season. Spring was the busiest time of the year, with multiple crops needing to be tended and picked before everything wound up in November. 'There were decisions to be made really quickly if we wanted to finish,' she says.

As Pam became fond of telling her daughter, the 'carousel is not going to stop for me to get off for a while; you just have to keep going'. To make matters worse, she found herself coping with a serious family illness. Ian's aged mother was diagnosed with kidney cancer. 'They were doing tests all the time but they didn't operate on her until the middle of November, and I pretty much moved in to look after her,' she says.

Born in Brisbane and mostly raised in Bowen, Pam didn't come from a farming background. If her life had taken a different path, she might have been a chartered accountant. 'I love book-work,' she admits. Instead, she trained as a banking clerk after leaving school, making the most of her exceptional aptitude for numbers. Employment conventions in the sector at that time meant she had to give up her career when she married, so she worked on the property, putting in long, hard hours. 'You marry a farmer, you've got to farm,' she says when asked if she liked her new life. 'I think you just do what you have to do.'

In her time with Ian, she became a vital part of his farm and so did Belinda, but Ian had been responsible for the spraying and watering programs, about which they knew only a little. Understanding their dilemma, Peter Reibel called in every day before starting work on his own property, and talked through with Belinda what she needed to do. 'He basically turned up every morning at six o'clock, picked me up and showed me how to do the watering, fertiliser injecting, all that kind of stuff, for quite a few weeks,' Belinda says.

Not everyone was so helpful. In a moment that stands out in family lore, a manager employed by Climate Capital Packers turned to Pam when she walked into his office not long after Ian's death and said: 'So what are you going to do now? Women can't farm.'

It was a pivotal moment for Pam, providing the motivation she needed to keep going, if for no other reason than to prove him wrong. *Just watch me!* she thought. Knowing Belinda's involvement was critical, she went straight out to the paddock and found her. 'Will you give me the next five years of your life?' Pam asked her daughter.

'What are you talking about?'

'Will you give me the next five years of your life?' Pam repeated, before explaining what the manager had just said. Responding in much the same way as her mother, Belinda did not hesitate—even for a moment. 'Game on!' she said.

Pam admits that every now and then she has wondered what might have happened if the manager had not said what he did, or Belinda had refused. She might have gone back to work in the bank, or perhaps set up her own business doing farm plans. However, she is absolutely certain that she wouldn't have kept farming if her daughter hadn't wanted to be part of it.

For the next six or seven years, Pam and Belinda forged on together, proving without question that women can indeed be successful farmers. Well-organised and methodical, Pam sat down within days and made two lists—one outlining what she and Belinda were good at, and the other identifying their weaknesses. Financial management and bookwork were definitely on the list of strengths. Another major asset was the people around her—Belinda and the employees on the farm. 'I never felt as if I achieved anything. It was the team behind me. I just told them what to do,' Pam says. 'All my team were good at something; it was probably more about what they could do without stretching them too far.'

In the end, she decided to hire an agronomist to provide technical advice on soil and plant nutrition, and controlling pests and diseases; and a contractor to handle chemical spraying. After some issues with the contractor, she bought a spray rig so they could do it themselves. The company trained one of the employees how to use it and when he left it became Belinda's job. 'He hopped in the spray rig with me one day, we did two laps, then he hopped down. See ya!'

'Her whole life has been sink or swim,' Michelle comments, hearing this story.

While Pam took responsibility for financial management, Berl became increasingly responsible for growing the produce. Like her mother, she recognised the value of the team around her, and how much she could learn from them. Two people in particular stand out from that time—Keith 'Hopper' Anderson and Greg Odger, her best friend's father. 'It wasn't necessarily about them teaching me stuff, but just watching them. I'm a watch-and-learn kinda person, that's how I pick things up, and what I've learnt from them over the years has been huge,' Belinda says.

Along the way, there were tough times financially, although Berl was not always aware exactly how tough. 'There were some very bad years in there, not making any money. I knew it was Struggle Street when prices were low, but I wasn't the one lookin' at the books. It was Mum's business then. I knew generally what was going on but I was out on the ground, managing staff and getting shit done.'

Pam soon came to the conclusion that the Collinsville Road block was more trouble than it was worth. It was too far away to supervise employees properly while also keeping an eye on the home farm, and too much time was spent travelling back and forth. Even before Ian's death they were beginning to realise the ten-year lease might have been a mistake, so Pam decided 1997 would be the last season there.

Then the 24-hectare property they were leasing at Delta became available to purchase. Ian had a 'gentleman's agreement' that he would be given first option, and the owner came to see Pam. The opportunity came up only twelve months after Ian's death, much sooner than she expected, but after deep consideration Pam agreed to buy it. 'It was a big decision,' she says.

In their first season on their own, Pam and Belinda grew watermelon and pumpkins. With their total vegetable-growing area now back to around 49 hectares, they decided to concentrate mainly on capsicums the following year, going on to specialise in producing miniature varieties, with export markets opening up in New Zealand and potential for sales into Hong Kong. Their produce was of such high quality it was used by a chef who cooked for Queen Elizabeth during her visit to New Zealand in 2001.

Berl was part of the small team that did all the tractor work to prepare the ground for each crop, and Pam spent a lot of time on the planter, putting in seedlings. At that stage the farm had a permanent staff of five, including a full-time mechanic, and about fifteen pickers in peak season. Back in the office, Pam kept an eagle eye on the overall operation and coordinated incoming orders, working with an employee at Climate Capital Packers, who was responsible for marketing their produce and sending it to Brisbane, Sydney and Melbourne wholesale markets, as well as national supermarket chains. It wasn't unusual to receive a last-minute phone call wanting more of a particular line, so the pickers would have to be quickly reorganised to fill the order.

Working intensively with Belinda throughout this time, Pam was careful not to push her into doing anything that she didn't feel genuinely passionate about. 'I'd always thought that when you get out of bed and say I've got this, this and this to do today, you can cope with whatever comes along. But when you get out of bed and go, "Oh God!", the first thing that goes wrong you are in the wrong frame of mind to deal with it,' she explains.

There was no question in Pam's mind that her daughter enjoyed growing fruit and vegetables but she had to 'nail her feet to the office floor' when it came to administrative tasks. Pam would always arrive at the farm office early. By the time Belinda showed up, she had a list of questions ready and Berl was not always receptive. 'I probably like too much detail and Belinda doesn't supply enough sometimes. I like to know everything. I had all these things ready to ask her, and I wanted the answers now.'

Belinda reckons her mother is very much like Pam's father in that regard. If he asked someone to do something and they agreed, he expected it to be done straight away. 'I'm a bit more laid-back,' she says. For her part, Berl admits that in the early years, she irritated her mother by constantly questioning why certain things couldn't be done differently. 'Because it's tried and tested, like reinventing the wheel,' Pam would tell her. 'But why?' a persistent Belinda would ask again.

Pam and Belinda soon found practical solutions to their different working and communication styles, which also improved the way they managed employees. About ten minutes before the end of each day's work, they would sit down with the two or three key staff responsible for supervising crews and talk through priorities for the next day and the rest of the week. They also set up a large whiteboard in the main shed, where everyone could see it. It identified priority jobs for the day, jobs for the next day, tasks that needed to be completed before the end of the week, and then things people should tackle when they had nothing else to do and couldn't find a supervisor to ask. Instead of standing around the office door in the morning wondering what was happening, they now had a clear idea of what was expected and could get on with it.

The result was not only a neat and tidy operation that drew comment from industry peers but a more efficient business achieved by paying attention to small savings.

'You can lose so much money and productivity if you don't pay attention to small things,' comments Michelle, who is in awe of Pam's detailed understanding of every single cost involved in producing every line grown on the property, down to the kilo. Even though she's now in her seventies and supposedly retired, she still does most of the bookwork and helps Belinda prepare the plans for what will be grown each season.

In the initial years after Ian's death, Pam applied that same acumen to Climate Capital Packers, whose partners were happy for her to stay involved in the business. In particular, she was responsible for overseeing the nursery division, which employed about a dozen people to grow seedlings for both the partners and other growers. Pam spent time there most days, checking what needed to be planted to fill orders and making sure the required seed was ordered in.

In 2002, she decided to leave the partnership and focus on farming with Belinda. Juggling all her responsibilities became too stressful for Pam when her mother fell ill. Ian's mother wasn't well again either. 'I don't know what I would have done if they hadn't got sick, but I'm just one of those people who never wants to say, "I wish I'd spent more time with them." That was why I stepped back.' Breaking the news to her partners, Pam decided not to ask the true market value for her share of the business because she didn't want to put any of them under excessive financial strain. 'I could have got a lot more for it, but I don't regret it,' she says.

Refocusing on their own growing operation, Pam asked Belinda what direction she wanted to take. A lot of the farm's storage and packing infrastructure had been dismantled after the partnership was formed because it provided access to central state-of-the-art facilities. 'What do you want to do?' Pam asked. 'Do you want to spend a lot of money and build bigger sheds and put in cold rooms or do you want to grow something that doesn't need that?'

The answer was surprising.

Being successful as a producer in the horticulture industry is about more than knowing how to grow fruit and vegetables. Markets can be highly volatile, depending on supply and demand, and the margins are often very small, with growers constantly under pressure from supermarket chains to keep prices down. Add to that the fact that most fresh produce has a short shelf life. Sometimes growers have only a matter of hours to pick it before the quality is affected and some lines don't keep long after harvest. Growers are often forced to take what they can get, or make the heartbreaking decision to leave their produce to rot rather than lose even more money harvesting it.

Mitigating these risks is one of the reasons growers keep an eye out for exclusive new product lines that offer greater financial security. For Pam and Belinda, their next 'big thing' proved to be a big thing. After years of experimenting, they pioneered growing large pumpkins for the emerging Halloween market in Australia. It all started in around 2000 when a seed company approached Pam to trial some varieties of what are known as carving pumpkins, grown in America not for eating but for carving into jack-o-lanterns for Halloween celebrations. Pumpkin carving is hugely popular in the United States, and the seed company was hopeful it might take off here too, given Australia's tendencies to follow American trends.

Always open to trying new things, Pam and Belinda agreed to give it a go. They began by planting no more than 100 seedlings to see how they performed in local conditions. 'You never jump in boots and all,' Pam warns.

The first lot didn't like the hotter conditions at Bowen so the seed company rep made a special trip to America to find alternative varieties better suited to the climate. In the second year, Pam and Belinda grew 500 plants to make a more thorough assessment. The pumpkins still proved more challenging to grow than mainstream varieties, not least because of the opposite seasonal pattern to North America, where Halloween occurs in autumn. However, Pam and Belinda were now confident they could produce them in viable quantities, so they engaged a marketer to find wholesale customers. By the time Pam decided to leave Climate Capital Packers and refocus on their own farm, the country's major supermarket chains were on board but the scale of production was still relatively modest while they kept playing with different varieties and growing practices.

It took seven years of patient trial and error before they settled on one particular variety that suited the Bowen climate perfectly and met the right specifications for the supermarkets. People entering giant pumpkin competitions might love to see gourds filling a wheelbarrow, but these needed to be easy to handle and affordable for customers paying by weight, while also large enough for carving and to bring a decent return. 'This is the juggling act,' says Pam. 'We thought they had to be big, but nobody wanted to go into a supermarket and get a huge pumpkin.'

They also had to fine-tune the harvesting process. Unlike popular eating varieties, carving pumpkins have thin skins and are mostly hollow inside. 'When you are picking them, you have to pretend you are picking up an egg or a baby. If you drop one, the inside comes away and they start rotting from the inside out,' Pam explains. 'And they have to be cured. You can't just pick them and put them straight in a bin. There's a lot of skill to it.'

Well aware of the potential for others to jump on board and flood the market they had invested so much in developing, the two women negotiated exclusive rights to their chosen variety, paying a royalty every year based on the tonnage grown. They even went to the extreme of never referring to the variety by its identifying catalogue number or giving it a permanent name. Even the commercial nursery growing the seedlings didn't know what it was. One year, Belinda might call it Spooky Pete and the next Scary Jill.

The market wasn't without its risks. If many horticultural lines are time sensitive because of their short shelf life, Halloween pumpkins are even trickier, given the market essentially revolves around one day of the year, 31 October. If the pumpkins mature too late, the market no longer exists. To spread this and other risks, and to reduce transport costs, Belinda signed up two growers to help produce the pumpkins—a friend in the Burdekin region and a grower in Broome in Western Australia's Kimberley.

By 2015, the crop had expanded to about 100,000 plants. By 2018, the three growers were producing 550 tonnes of

Halloween pumpkins, with Belinda selling her produce under the Stackelroth Farms label she had established in 2007 to honour Stacky. Supermarkets across Australia were taking them, as well as produce markets in almost every capital city. To help drive this expansion, Michelle and a very reluctant Belinda hit the publicity trail. Media were fascinated not just by the pumpkins but by the idea that the principal growers were women. Michelle even appeared on Channel 10's prime-time television program, *The Project*.

There wasn't much need for marketing in 2020, when the pandemic saw an explosion in demand from families trying to keep children occupied during lockdowns and travel restrictions, particularly in Melbourne. The demand was so high in Victoria that stock sold out very quickly and pumpkins had to be diverted from Queensland stores. Even though she was retired, Pam came to the farm every morning during harvest, monitoring the amount of produce being harvested and brought back to the sheds, then, at the end of every day, calculating the total cost per kilo to grow, pick and pack the pumpkins.

Belinda was preparing to do it all again in 2021 and even expand production given the ongoing pandemic, when she realised it just wasn't in her. It wasn't just the workload or an ongoing labour shortage, but the intense pressure involved in business negotiations to secure a price for the produce. 'I had my farm plans drawn up and proceeded not to sleep for about ten days. Michelle knew I wasn't sleeping and I was angry,' she explains.

Realising she had an important decision to make, Berl drew on the calm experience of long-time friend and farm-hand, Carolyn Kingston-Lee, known as Caz, who had sensed something was bothering her. 'I bounce things off Mum and Michelle, especially financials and big things, but Caz is like my right-hand person on the ground here. I feel like I'm not really her boss because we work together so well. There are a lot of times I go, "What do you think? Come and have a look." She is with us every day and she's seen the pressure through Halloween. So we sat on the table in the packing shed for like four hours, just throwing things backwards and forwards.'

Towards the end of the conversation, Caz said: 'The decision is ultimately yours, but since we've been talking I think you've made it because your attitude has already changed. The last week you haven't been yourself.'

Mind made up, Belinda rang her mother. 'I was waiting for this phone call,' Pam told her.

Stackelroth Farms still grow a few Halloween pumpkins, but since 2021 the focus has been growing less lines in smaller quantities, for direct sale through independent local businesses and their own farm shop.

It was Michelle's daughter, Stevi, who pioneered the farm's move into having its own sales outlet, when she was in Year 2. Encouraged by Michelle, who saw it as a way for her daughter to learn about money and improve her numeracy skills, Stevi set up a little wooden bin in front of the house. She filled it with butternut pumpkins that she had grown herself, and kept the proceeds as pocket money.

After Stevi left home, Michelle and Belinda thought it would be a shame to let the idea go so they repurposed an old wooden farm cart and placed it next to the main farm entrance. Then when the fruit and veg outlet on the main road closed, they ramped things up further so that the 40 or so families living in the Delta area could still buy fresh produce locally. They began by supporting other local growers, inviting them to supply produce, then topped it up with additional lines brought in through their longstanding Brisbane agent. The stall did so well that Belinda and Michelle had a shipping container converted into a shop, which has also become popular with passing grey nomads, who are particularly fond of small packs Caz puts together, with a selection of either fruit or vegetables.

~

It's Saturday afternoon and things are chaotic in the undercover area where Stackelroth Farms produce is being packed. Set on a raised platform built at one end of a large shed, it's only a small space but Berl likes it because it's convenient to the house and the roadside shop. The main cook, she can easily pop inside to put dinner on, and then come back and continue working. If she has time, Berl can even sit on the steps to watch the sun set behind the hills stretching across the western horizon. Looking out across some of her paddocks, it's her favourite place on the farm. She's not too sure of all the names but Mount Roundback is the tallest at 709 metres above sea level, then there's Mount Pring, which is less than half the height, and the even smaller Summer Hill. 'Grandad used to say if clouds were sitting on all three it's gunna rain,' Belinda says.

The clouds are too high for that today, but the weather bureau is forecasting a deluge early in the coming week, with up to 1000 millimetres predicted for parts of the Queensland coast and 500 millimetres for the Bowen area. The skies are already heavy and grey, and the air has been thick with unseasonal humidity for days.

Prone to not sleeping well at the best of times, Belinda was awake most of the night worrying about everything she needs to get done before the heavens open. If the forecast proves accurate, most of the farm will disappear under water so she's dumped plans to plant zucchinis on some lower ground. Even though it looks flat to the human eye, her paddocks slope slightly towards the Don River, about 1.5 kilometres away. After heavy rain, it becomes one of the fastest flowing rivers in the Southern Hemisphere, and the spillover has flooded the area many times.

Then there's the threat of fungal diseases damaging the zucchini and tomato plants Berl has just started to pick. She's waiting on a call from her agronomist to discuss a preventative approach that is likely to involve applying fungicides before the rain starts. And she wants to add a second string to some of the tomato trellises, pulling the vines up higher. Meanwhile, there is the farm shop to tend. Belinda opened the doors at seven o'clock and a steady stream of weekend customers is rapidly emptying the shelves. That means more produce has to be collected from the cold store and taken down in the farm buggy. Some of it has to be packaged and labelled too.

Adding to Belinda's stress levels is the prospect of crowds coming to the farm tomorrow for a special event Michelle is organising. It will be Mother's Day and she is planning family activities on their front lawn, with bees and honey-spinning as the focal point. To help promote the day and add to the attraction, Michelle is selling sunflowers for people to give to the mothers in their lives.

Adept at using social media to market their produce and promote community causes, she has posted a photo of 'The Boss Lady' holding a bunch. Berl is not keen on this aspect of the business but over the years she has come to recognise the importance of connecting with consumers, if they are to appreciate the importance of eating fresh, healthy food and understand how it is grown.

'However you feel comfortable my love!' Michelle instructed dryly during the brief photo shoot, earning a glare from her reluctant partner.

'What d'ya say?' Michelle teased, once the photo was taken.

'Happy wife, happy life!' Belinda responded dutifully.

'Yeah, ya do!'

Concerned the Mother's Day event might be rained out and leave them with unsold flowers, Michelle's post encouraged online orders and gave customers the chance to collect them in advance, 'because life isn't crazy enough!' The response has been overwhelming, so she is compiling a list of sales to make sure there are enough bunches. It's taking up time she was meant to be spending going into Bowen to pick up last-minute supplies for tomorrow. She also needs to meet some volunteers at the Police-Citizens Youth Club (PCYC), which she manages. Then there are the signs she has to paint and the bee frames she needs to pull from her own hives so they are ready for the honey-spinning demonstrations.

Michelle knows some of these tasks won't get done. 'I'm an 80 per center,' she confesses. A problem-solver and creative by nature, her mind is always racing ahead to the next challenge rather than finishing what she has already started, which infuriates Berl at times.

She is still working her way through the list of sunflower sales, when a convoy of vehicles turns into the driveway leading to the shed. It's friends they made while crewing for the annual Variety Jet Trek, which is held on Australia's east coast to raise money for charity. They arrived in town the previous night on a personal adventure, riding jet skis from Mackay to Cooktown. One of them is well-known business entrepreneur Christine Taylor, who turned a small dog-washing venture she started at the age of sixteen into a multi-million-dollar business with franchises around the world washing 20,000 dogs a month. 'I could never do that,' whispers Ella,[2] a shy fifteen-year-old helping to pack vegetables.

With a challenging home life and lacking in confidence, Ella has been taken under Michelle's wing after coming to her attention at the PCYC, where she attends a flexible learning centre set up to support disadvantaged youth. Detecting a spark of something in the girl, Michelle has offered her a few hours' paid work so she can spend some one-on-one time with her doing some mentoring. Never one to let an opportunity go by, Michelle introduces Chris and shares something of her inspiring story, pointing out what she has achieved 'just from trying something new and backing herself'.

Nurturing potential in young people is not just part of Michelle's role as manager of the PCYC in Bowen; it's also driven by her own traumatic experiences which she talks about openly as a way of connecting. Although she makes light of it, it is not an easy story to tell.

---

2 Not her real name

Born in Hughenden, Michelle was initially raised as one of four children, although she has since discovered that she has at least thirteen half-brothers and sisters, with all of them sharing the same biological father. 'I found out I was one of five by the age of twelve. Then in Grade 9 I found out I was one of twelve, and now I'm one of seventeen. It's batshit crazy—you couldn't write a movie about it!'

Her parents had a troubled relationship, which undoubtedly left mental scars on their children. Michelle recalls hiding under the bed listening to them fight and being flogged with the cord from an electric iron. 'I think there are other things that I've blocked,' she says, explaining that her coping mechanism is to surround herself with chaos and white noise so she doesn't have to deal with the past. 'I fill my life with busyness because it's a distraction.'

When Michelle was five her mother walked out, taking one child with her and leaving Michelle and two other siblings with their father. Six months later he died of cancer. Authorities stepped in and placed the children in emergency foster care. They changed homes a few times before they were taken in by Marie and Len Kirkland. Marie was familiar with their circumstances because she worked at the Bowen Hospital and had nursed their father in his final days.

Michelle remembers travelling to the Kirklands' house with an apple box sitting on her lap, containing everything that she owned. Badly neglected for some time, she was seriously malnourished to the point where her fingernails were like thin layers of rice paper, and her head was infested with lice and nits. To solve the latter problem, Marie cut Michelle's long, thin red hair short, like a boy's. 'She wasn't a feminine woman but I loved her to death,' Michelle says, comparing her to Matron Sloan in the old television series, *A Country Practice*, who was a stern disciplinarian with a hidden softness.

When Michelle was about seven, Marie and Len split up. 'I remember sitting in the lounge room when she told me and shaking uncontrollably 'cos everyone that I loved had left, died or whatever, but they stayed friends, very respectful of each other, and they stayed married even though they didn't live together. They were still Mum and Dad.'

While Michelle's siblings went to boarding school, she remained home to complete primary school. Effectively a single parent, Marie continued working shifts as a nurse so they had money to live on. 'I was a latchkey kid. Sometimes we wouldn't see each other for a week, but we'd leave notes for each other. I would come home and let myself into the house and sit in front of the TV and eat a litre or two of ice-cream, without even realising I'd done that, or dinner was in the microwave, and it was saveloys and cheese and onion. But Marie did the best she could. I would not be the person I am without her. She was my saviour.'

After finishing high school, Michelle found employment at a local fast-food chain. She worked for the same chain in Yeppoon when she moved there just before her eighteenth birthday. A year later she upped sticks again and went to Rockhampton, where she gained experience in various businesses, including management roles at a shoe shop, the local store of a well-known fashion brand, and Lifeline's regional distribution warehouse. Michelle even spent time pulling beers at the iconic Great Western Hotel. Often holding down two or three part-time jobs at a time, she also sold flowers in night clubs for quite a few years.

At the age of 25, after something of a whirlwind romance, Michelle married a musician who worked for the postal service. 'He was a nice guy but it was always Struggle Street for us,' she says. Their financial situation was not helped when he gave up regular work as a postman to focus on his music and play full-time with another musician. Then came the day his colleague decided he didn't want to continue. They were setting up for a gig in Rockhampton when her husband broke the news. Michelle was seven months pregnant at the time.

Wanting to reflect on the situation they were now facing, she set off alone to walk home, which was quite a distance. Determined their child would have a secure start in life, Michelle realised she would have to take responsibility. By the time she reached her front door, she had decided to join the Queensland Police Service (QPS). It wasn't a completely random idea; she had considered it as a possible career when she was in high school, along with becoming a teacher. Pondering her options now, she realised teaching wasn't realistic, given it would involve four years of juggling a baby while studying before bringing in a salary. However, joining the police service meant only six months of training, and she would be paid while doing it.

As soon as she got home, Michelle contacted the QPS recruitment office for information about enrolling, and then rang a friend in the service to ask their advice. She discovered the QPS was keen to sign up mature-age candidates with some life experience, as long as they could demonstrate the capacity to study. Undaunted, Michelle registered for a Diploma of Justice course at the local TAFE.

Her daughter, Stevi-Leigh, was just shy of two years old when Michelle entered the police academy at Townsville in 2001, at the age of 28. After completing the regulation six-month training program for new recruits, she was assigned to her old home town of Bowen for constable training, and has never left. Michelle was happy to be back in the town where she grew up, however her marriage was going downhill 'at a great rate of knots' and soon ended. Instead, she found joy in reconnecting with old schoolmates.

Even though they were the same age and had played in the same basketball team, she and Berl had different circles of friends at high school and didn't know each other all that well. That changed when Berl's best friend, Jodie, recognised Michelle's face when it appeared in the local newspaper. The three women began hanging out socially, and then Michelle started helping Belinda at the farm after work. 'There was absolutely nothing in it,' she explains. 'Then Belinda and I just started getting closer and closer.'

One of the first things that attracted Michelle to Berl was her hands. She brings up an image kept on her phone. It shows one of Belinda's hands in close-up—lined and work-worn, grimy with soil and holding a ripe red tomato, which she is plucking from its vine. 'Doesn't that speak volumes? It's the epitome of Belinda. The roughness and the strength, but the gentleness.'

The two women were 30 when their relationship started. It took about eight months for them to move in together and a lot of soul-searching. Michelle came with an almost five-year-old and she needed to be certain it was the right decision for her daughter too. From her perspective, Belinda had never wanted children and claims not to be very good with them, although Michelle says that definitely isn't true.

For her part, Michelle soon realised taking on Berl meant taking on the farm, and giving up her own career ambitions. 'I was going to go out west, and move up the ranks. I was going to be an inspector. But Belinda was never leaving Bowen, so I had to sit back and think about what I wanted in life, and I reckoned our relationship was worth investing in.'

However, they remained careful around Stevi for some time. 'She would never see us walk into the same bedroom together, or go to bed at the same time . . . because I wanted to make sure that what we were doing was secure. For Stevi, Berl was just this chick we lived with.'

It took quite a while to make their relationship public too, which kept curious friends and neighbours guessing. 'Can't two friends live together?' Michelle would throw back at anyone who asked. She was very conscious of the potential impact, given she was a police officer, fairly early in her career, and working and living in a small community. She also didn't want to jeopardise custody arrangements for her daughter. 'It was huge for me . . . so we were sort of undercover for a while.'

Then one night at a gathering of friends to watch a State of Origin game, with speculation rife, one of their neighbours came up and said: 'Seriously, if you two are on together would you just slap each other on the butt or something, just to put these people out of their misery. We love you, whether you are or you aren't!'

There wasn't much backlash once word spread, and Stevi's father was cooperative too, when it came to working out arrangements for his daughter. Talking about the Bowen community and their network of friends, Michelle says: 'We are very fortunate. People just see us as Belinda and Michelle—they just see us for us, and our authenticity.'

Adjusting to life together wasn't always easy. Michelle was anxious that Belinda not feel pressured to look after Stevi in the early days of their relationship, and there was a farm to run, so Michelle organised home day care. But working shifts and being on call for emergencies was part of her job, so Belinda was often left caring for Stevi. There were many nights when Michelle worked long after her shift was meant to end, coming home hours after her daughter had gone to bed. And given the way Stacky had died, Belinda found it impossible not to worry if Michelle forgot to let her know she would be late.

In 2011, Michelle found her metier when she was appointed manager of a brand-new PCYC opened with great fanfare after three years of community fundraising. Bowen was the 50th club established by the Queensland Police Service to provide youth and community programs that focus on personal development. The clubs are mostly located in low socio-economic areas where many of the young people they support are at risk or come from disadvantaged backgrounds.

The state-of-the-art Bowen facility was built at a cost of around $5 million, with funds from the state government, Whitsunday Regional Council and mining company Xstrata (now known as Glencore). Being placed in charge was quite a feather in Michelle's cap. She had no particular training in youth work, but she figured her own personal experiences combined with the years she had spent in various businesses would stand her in good stead.

Even though it's located on the western edge of Bowen, overlooking Denison Park, the building in many ways lies at the heart of the community. After operating for only seven months, it had almost a thousand members in a town with a resident population of around ten thousand, and had notched up more than twenty thousand attendances by people keen to take advantage of the programs and facilities.

The building incorporates a multipurpose gymnastic stadium that would be the envy of towns twice the size, three full-size squash courts, and a large function room that is hired out for conferences and meetings. People come there to access everything from after-school care and kinder-gym sessions, volleyball, basketball, boxing and dance classes, to crime prevention and drug-awareness programs. While there is a team of about ten staff, volunteers play a big role in delivering initiatives such as Braking the Cycle. The award-winning driver mentor program links experienced drivers with young people who don't have access to a vehicle or someone who can supervise them, so they can complete the required number of logbook hours to get their licence.

Managing the PCYC can be all-consuming. Michelle describes it as her addiction because 'it feeds my own self-worth'. At times, it has placed enormous strain on her own health, as well as her personal relationships. The job may have taken Michelle off the road and away from having to work shifts, but it wasn't unusual for her to put in double the hours she was officially paid to work. Despite being exhausted, she would often lie awake at night, worrying about how to make sure there were enough funds to provide job security for its part-time staff and keep essential community programs going.

'This place is both a joy and a curse because it was built for the community from government money and council money, but we only got the shell. Everyone thinks we have a big budget and we get government money to run it but we don't. The Queensland government provides the wage for this person here, and then this person has to make it work, so I'm the business manager, I'm developing all the future projects, activities, I'm the youth worker, I'm crime prevention. To do it properly you need 90 hours a week and I was doing up to 90 hours a week at times, for five years.'

Then she crashed. Completely burnt out, Michelle took six months' long-service leave. 'I was tears and tantrums,' she says, admitting to being terrified once she realised the strain it was placing on her relationship. These days, she has learnt to step back a little, claiming she has mellowed. Most days she starts at nine o'clock, but she will often still work until seven or eight in the evening before coming home to a hot meal.

After ten years, Michelle still loves the community-engagement side of her role. Even before she started managing the club, it wasn't unusual for her to call into a school or childcare centre and say hello to the children, so they weren't only encountering a police officer in times of trouble. 'I joke I'm the highest paid bingo caller, face painter, bus driver in Bowen. But if you see a photo on the PCYC Facebook page and I'm sitting there in full kit, in my sergeant's uniform, face painting kids at a childcare centre, there is value in that.'

Working with the town's youth and creating a supportive environment for them, has given Michelle a sense of life turning full circle. Because of her own experiences, she is well aware of the difference it can make when others take an interest. The truth of this came home a few years ago when she was speaking to a community gathering about the PCYC and why its work is so important. 'Society and educators tell us to invest in your future, and for me investing in your future is not about saving to buy a house or a car, or anything like that. Investing in our future should be about investing in our young people, because they are our future,' she told them.

To bring the point home, Michelle painted a picture of her own rocky start in life without telling the audience, at first, who she was talking about. 'It was the grace, compassion and care of people in this community that changed this young person's future. Some would say that this person turned out to be a community leader,' she said.

Then she looked up from her speech notes. That is when she spotted Pam sitting in front of her and someone who had been on the Catholic primary-school board that agreed to let Michelle and her siblings attend the school for free, when they heard about their family situation. These people had made a significant difference in her life. Realising the proof of what she was saying was right there in front of her, a lump formed in her throat and she fought back tears to deliver a punchline that astonished many in the room. 'I can tell you this story because it's my story.'

Recounting this tale, Michelle says that many of the people who had stepped up over the years to help her had no idea at the time of the potential impact of their seemingly small acts of kindness. 'You can change things in a generation with the right support,' she adds with total conviction.

~

Michelle had only been back from long-service leave for three months when Cyclone Debbie hit in 2017. Initial tracking of the tropical low-pressure system had it making landfall around Townsville and missing the Whitsunday region, but over the next few days that changed. By the afternoon of Monday, 27 March, forecasters were predicting it would hit the mainland just south of Bowen the following morning, and might even register Category 5 on the Tropical Cyclone Intensity Scale—the most severe rating possible. There were additional fears the cyclone would coincide with high tide, causing massive surges of seawater to inundate low-lying areas, so thousands of residents and business owners were urged to evacuate.

Emergency services were also concerned that most of the town's homes were at considerable risk of destruction because they had been constructed before stricter building codes for cyclone resistance were introduced in 1985. Police and SES leaders implored people to leave, reminding them that the town's cyclone shelter at the state high school could only hold around eight hundred people and should be considered a place of 'last resort'.

Belinda had been suspicious for days that the initial tracking was wrong. *This is the one that is going to hit us*, she thought. Working day and night, she moved as much as she could into the workshop and packing sheds. Trucks, the tractor and other equipment were parked up against timber bulk bins to hold them secure, and the spray rig filled with water to weigh it down. Pallets were gathered up and ratchet-strapped to truck trays; and all the paperwork and files in the farm office were bagged and placed in plastic tubs. Joining the rush on local supermarkets and petrol stations, Belinda also made sure to stock up on cartons of water and fill every available container with fuel for her generators.

Working alongside her was Stevi—at seventeen, already a veteran of several tropical cyclones. As a young child, she'd watched through a glass-topped door while her mother and Belinda tried to tie down the roof of a small shed in their backyard. With every gust of wind, the roof lifted, battens and all, and then crashed down onto the walls—bang, bang, bang. If it came completely loose, odds were that it would smash into powerlines running overhead and bring them down on the house.

Even though it was risky, Belinda had rushed outside and belted a series of star pickets into the ground, then grabbed some fencing wire. Once everything was lined up, she called Michelle over and got her to hang on to the roof while she tightened the wire. Meanwhile, Stevi kept watch from inside, mobile phone at the ready in case something went wrong. 'All I can remember is turning around and seeing this poor kid just looking through the window, and her eyes. And I just thought, "Holy shit, what have we put her through!"' Michelle recalls. In testament to the moment, the star pickets are still there.

That cyclone had been scary enough but meteorologists were predicting Debbie would be much worse. As Monday rolled on, heavy rain and strong winds began to lash Delta.

Belinda, Michelle and Stevi spent that night at Pam's house, where they planned to wait the cyclone out. Joining them and Pam were five dogs, and Michelle's foster mother, Marie, who was by then a resident in the Bowen aged-care home. The nursing home was in a flood-prone area and Pam's house had been renovated some years before to a high standard of resistance to cyclonic winds, so they figured it was a safer place to be. They reassured a police patrol that dropped in around daybreak the next day that they were well prepared.

Cyclone Debbie made landfall near Airlie Beach, about 80 kilometres south-east of Bowen, at 12.40 pm on Tuesday, 28 March. At its most severe, it rated Category 4 and generated a peak wind gust of more than 260 kilometres per hour—the strongest ever recorded in Queensland. To make matters worse, the weather system was slow moving, creeping along at just 7 kilometres an hour, exposing the areas it hit to sustained destructive winds. The town of Airlie Beach was torn apart and so was Proserpine and resorts in the Whitsunday Islands.

The cyclone finally weakened in the early hours of 29 March, but it kept producing damaging winds and torrential rain that stretched from central Queensland to the south-east, with several locations receiving up to 1000 millimetres in two days. Not known for hyperbole, the Bureau of Meteorology described the rainfall as 'phenomenal'. Some places received the equivalent of half their annual average in just a few days. So much water came down the Don River that it rose more than 5.5 metres at the Bowen pump station just before midnight on the Tuesday. At Pam's house, the noise outside was horrendous while inside it was barely discernible, although water forced its way in around the edges of the windows and doors.

Once the cyclone passed, Belinda and Michelle went to check on the farm. They discovered an old shed had been destroyed; its corrugated-iron sheeting was spread across the paddocks. Their home hadn't fared well either. Tradesmen had finished installing a new kitchen just two days before. Now they were facing a substantial rebuild that would involve gutting half the house and replacing internal walls as well as the roof. With the farm coming first, it would take fourteen months for this work to even start. Meanwhile they lived in what had been the spare room and showered in the back shed, while Stevi slept in a campervan parked in the backyard.

Leaving Belinda to work out where to start the clean-up, and feeling terrible that she couldn't stay and help, Michelle headed into the PCYC, where she discovered water inundation had severely damaged the gymnastic stadium and squash courts.

That night the Bowen area was battered again by terrible thunderstorms and torrential rain, which Belinda says did more damage than the actual cyclone. Driving back from town, Michelle reached Pam's house and was too alarmed to leave her vehicle. 'The electrical storm was just nuts. I sat in the car for well over half an hour. I've never seen anything like it.'

Anxious to stay alert, Belinda made a bed on the floor in the lounge room that night, leaving Michelle to get as much sleep as she could in her old bedroom. Michelle recalls lying there and hearing the end of a branch from a big black-bean tree scrape down the window after snapping off. A couple of other trees came crashing down, barely missing the house.

The next few days were an exhausting blur of hard physical toil and intense emotion, heightened by the loss of a beloved pet, a border collie call Suey. 'We knew she was getting to the end of her life, but the cyclone took it out of her,' Belinda says. None of the Bowen vets were open so she and Michelle drove the dog more than 100 kilometres to Ayr, where she had to be put down.

Back at the PCYC, Michelle pulled her small team together, deciding it was important in a time of crisis to keep the doors open and do what they could to help the community, even though the building was damaged. For weeks to come they operated out of the intact side of the facility, continuing to run key services such as after-school care. The club also became a central point for people to drop off donated items such as fridges, bedding and clothing. Michelle and her staff then helped distribute them to where they were most needed. 'I had a really awesome team,' she says proudly.

Meanwhile, out on the farm Belinda and Stevi trudged through the mud and started picking up debris. The biggest concern was the black plastic that had been laid in ten 230-metre rows across 12 hectares of paddock, to nurture 30,000 butternut-pumpkin seedlings. The crop was a write-off and all the plastic needed to be pulled up by hand and removed, before the ground dried out and cemented it in. On another part of the farm, agronomist Chris Monsour was facing the same problem, with acres of plastic laid for some trials he was conducting.

Every day, Belinda dragged herself out of bed at about six o'clock and headed to the farm, returning to Pam's house covered in mud and exhausted by about three o'clock. After a cold shower, she went outside again with a chainsaw and cleared up fallen trees and branches. They had a generator and plenty of fuel to run it, so they could turn on some lights in the house and keep the fridge cool. A gravity-fed tank meant there was running water but there wasn't enough power to heat it and cooking had to be done outside on a barbecue.

The first Saturday after the storms, Michelle woke up crying, overwhelmed by the long hours she had already worked and worried about Berl, and the task that lay ahead cleaning up the farm and getting it back into production. Taking time off from the PCYC, Michelle donned her farm clothes and joined Belinda, Stevi, Chris Monsour and a couple of his employees in the paddock to pull out yet more plastic.

They had been slogging away all morning, when they looked over to see a group of people walking towards them. It was PCYC community development officer, Madonna McLeod, and a host of teenagers she had rounded up from the club's youth group—the RUBY Crew (the acronym stands for Representing and Uniting Bowen Youth). Years later, Michelle's voice still chokes up when she recalls the moment.

'For a day and a half, this tribe of awesome young people slogged their arses off. These beautiful people saved our sanity, and even with the hard work they smiled all the way through. It was just so uplifting.' Pointing at a photo of the clean-up team, filthy with mud, huge grins splitting their grimy faces, she adds: 'I get very irate when people say that young people don't do anything. You give them a purpose and they have the biggest hearts. That's our future there.'

As the weekend rolled on more people rallied to help—an English backpacker, Steve, who had picked for them two years before and was now living in Townsville, drove down with his fiancée. Police checkpoints were limiting traffic coming into the area, with flooding cutting some roads off, but Steve explained they were heading to help clear up the farm and they were let through. Madonna's husband and their youngest daughter showed up too, and so did the mother of a RUBY Crew member, even though she had a broken arm. When another helper couldn't come back to keep working on the Sunday, she made packed lunches for everyone and dropped them off instead.

By the end of the weekend all of the damaged plastic had been removed and Belinda was ready to start discing the paddock so she could plant a replacement crop. Fortunately, she had some seedlings left over, but there was no getting away from the lost income. The cost of cyclone insurance for horticultural crops is prohibitive so most growers don't have it, and Belinda had sunk all the profit made from the previous year's butternut pumpkin crop into planting the next one. 'We lost every cent,' says Michelle.

In need of a high-value crop that would help repair their finances, Belinda interplanted a mix of bite-sized tomato varieties for the first time that season and offered them as medley punnets. Production has since grown to more than seven tonnes a year, mostly sold through local markets. The ever-reliable Caz is chief picker, spending five days a week among the vines during peak season. She thrives on the solitude, listening to music while she carefully harvests the delicate fruit. 'I love it out here. I have my radio. I solve the world's problems. "Me time", I call it.'

Described by Pam as the backbone of the farm, Caz has been working there for more than a dozen years. She was raised on a nearby property that grew mostly tomatoes, capsicums and cucumbers. Passionate about bugs, she worked in agricultural research for five years, monitoring crops for both beneficial and harmful insects, and breeding ladybirds or lady beetles, which are natural predators of some of the worst crop pests, particularly aphids. She's tried living somewhere else a few times, but says Bowen is a vortex that just sucks people back.

Having the confidence of both Pam and Belinda, Caz often takes responsibility for supervising employees. According to Pam, she has the biggest heart and lots of patience, although Caz reckons it's tried every season when at least one male backpacker rocks up obviously thinking the women need a man to show them better ways to do things. These days the first question that she asks male employees is: 'Do you have problems taking instructions from women?'

During the 2021 season, Caz was mightily impressed by a crew from Tonga, who stepped in to pick when there was a shortage of backpackers because of the pandemic. The men were in Australia under a seasonal-workers scheme, and the grower who brought them over had an unexpected lull in work. Within four days, the crew of eight harvested 100 tonnes of pumpkins, compared with 10 to 15 tonnes per day in previous seasons, when the farm employed around twenty backpackers. With time to spare, Belinda put them to work in the packing shed, where Caz was in charge. 'They did the best pack-outs I've ever seen. They were absolutely fantastic.'

Caz says Stackelroth Farms has a good reputation for the way its workers are treated, and giving young people a start in the industry, employing them before school and at weekends.

'Belinda and her mum are just champions. Belinda is very humble, doesn't like the limelight of any sort, but Michelle pushes her a bit and that's good. She needs to see what she's done.'

Michelle is in her element. It is Mother's Day and, despite unsettled weather, a crowd of people has begun gathering on the front lawn at Stackelroth Farms well before the advertised start time. 'This is one of the first events like this we've put on. It's a little bit chaotic but we're getting there,' Michelle apologises to the crowd as she starts her presentation.

In many ways, organising the event is an act of compensation for Michelle, given her childhood. 'I had a foster mum, but celebrating Mother's Day was bittersweet, so we are doing this event so I can see other families happy, and I get to share in that. So that's my way of celebrating—making an experience for other people. I have siblings who would just sit and dwell, but I have to make the best of things. "If it is to be, it is up to me"—that's my motto.'

The audience is mostly young families drawn by a fascination with bees, which are the central theme of the day. Michelle seizes the opportunity to tell them something of the farm's history, before talking about bees and their importance in the food chain. In the past, Belinda has paid people to bring in hives to pollinate her cucurbit crops, which don't self-pollinate, but Michelle now has her own bees. She became interested in beekeeping about five years ago, and then Belinda bought her some hive boxes for Christmas.

Michelle has removed several frames of honeycomb and placed them in a container ready for today's demonstration. Part way through her talk, Belinda shows up wearing a baseball cap pulled low over her eyes and a bright blue shirt sporting the Stackelroth Farms logo. She removes a frame from the container and shows it to the children, leaving Michelle to quiz them on what they know about bees. They prove very knowledgeable for their ages, hands raised eagerly to get her attention.

With the main part of Michelle's talk done, the children crowd around Belinda to watch her decap a frame. Using a special tool with sharp, fork-like prongs, she removes the beeswax sealing the honeycomb, which starts oozing thick, golden honey. Lifting the frame carefully, she places it in the honey spinner standing nearby on a small wooden platform. Now it's time for the magic!

Unlike the conventional stainless-steel honey extractors used by commercial beekeepers, this model has bright red legs and a clear perspex body. Michelle bought it on a whim, even before she had her first hive, after seeing it advertised online.

'What d'ya want that for?' Belinda queried, trying to curb Michelle's tendency to be impulsive.

''Cos its cool and you can see what's happening,' she replied.

Looking at Belinda now, she teases, 'Aren't we glad we bought it?'

Belinda pulls a face, and the audience laughs.

Urging the children to take a step back so everyone can see, Berl chooses a volunteer to turn the spinner's bright yellow handle. The audience is soon 'oohing' with delight as the frame rotates with increasing speed, and centrifugal force flings the honey out of the frame onto the perspex. The honey coats the sides of the spinner, then slowly trickles down to an outlet poised above a stainless-steel sieve and large plastic bucket.

For the next fifteen minutes or so children take turns winding the handle, while their parents watch. Other families relax on blankets spread on the lawn. Under a small marquee, two members of the RUBY Crew are offering face painting. A second marquee is providing shade for several friends who have come along to manage sunflower sales. Among them is Sarah Bon, who went to the same high school as Belinda. 'I was this skinny little runt of a kid and she sort of took me under her wing,' Sarah explains. A couple of years younger than Belinda, she was being bullied by another girl. 'Berl is not a fighter or anything like that, she's too nice for that, but she can be intimidating. She just stood behind me and looked at this girl and she never bothered me again.'

Sarah admires Belinda for her loyalty and work ethic. She is a huge fan of Michelle's too, recalling the role she played in helping Sarah cope with a family tragedy. In 2008, her nephew Peter attempted to take his own life. He failed but was left in a vegetative state until he died in 2014. Determined that something positive should come out of such a terrible situation, Sarah set up The Peter Project ten years after his suicide attempt. The initiative involved taking on physical challenges such as long-distance walks, as a way of starting conversations and raising awareness about mental health.

Sarah was struggling to get The Peter Project off the ground when Michelle spotted her crossing the street in Bowen one day, head down and clearly dejected. 'I was feeling very defeated at one point because I couldn't afford the insurance to do what I wanted to do,' Sarah says.

She explained the problem to Michelle, who gave her a hug and then made some calls. Over the coming weeks, Michelle provided valuable advice and leveraged her networks to build support. Sarah began by walking from Bowen to Collinsville, a distance of about 85 kilometres. Next she tackled almost 500 kilometres from Bowen to Cairns, accompanied by her father, then the infamous Kokoda Track in Papua New Guinea.

'Berl was always in the background, helping make things happen too,' Sarah says. 'They are two awesome people, good for the community.' Pausing to look around at people enjoying the Mother's Day event, children playing on the grass, adults chatting and mothers holding bunches of carefully wrapped sunflowers, she adds: 'This is them.'

In reality, Belinda and Michelle have been combining their two working worlds for some time. It began around fifteen years ago, when a local Indigenous organisation, Girudala, brought some children out to Stackelroth Farms for a visit. There have been many farm tours and workshops since, with the two women passionate about educating people of all ages about farming and how their food is grown.

In 2016, Michelle took things a step further and roped Berl, Caz and Chris Monsour into delivering a PCYC project. Managed by Madonna, it encouraged disadvantaged youth to re-engage with education and develop skills that would open up employment opportunities in the horticultural sector. Participants came out to the farm for hands-on training to learn the basics of growing vegetables, while other sessions connected them with businesses in the horticulture sector that might offer future employment. The success rate for participants either finding work or going back to school was 98 per cent. Much to everyone's shock, the project was awarded top honours in the 2017 Premier's Awards for Excellence in the public sector. A stunned Belinda joined Michelle and Chris on stage at a gala event in Brisbane to collect the award. 'We were just going for a bit of a junket, and didn't for a minute think it would win,' jokes Berl.

In more recent years, as Farmer Belinda and Sergeant Michelle, the two women have collaborated with local libraries to run sessions for young children around the theme of bees and honey, in support of Michelle's role as a community literacy champion. At their Mother's Day event, Michelle met the parents of a young girl from Collinsville. They told her their child was so inspired after one of the sessions that all she wanted on her sandwiches for months was honey. Then there is the three-year-old boy with multiple food allergies, who came to the farm with his family to buy fresh produce. When he spotted Berl in the paddock, he ran towards her yelling, 'I need honey!'

'These are the experiences kids remember when they grow up, and Belinda is such a strong role model. Little kids just love her,' says Michelle.

Bone weary but satisfied after a long and successful day, Michelle is back in the house watching Berl dish up a quick dinner of mashed potato, pan-fried sausages and gravy. Oddly, given the plentiful supplies outside the door, there are no other vegetables. 'The Irish in me is lovin' you tonight, babe,' Michelle tells her.

Behind them, three dogs are sprawled out over a giant modular sofa. Harry is a rescue dog who showed up at the farm as a puppy and never left, and so is Buddy. He was found abandoned and full of mange just before the cyclone hit. There were no plans to keep him, but chaos reigned at local animal rescue services after the disaster and by the time the situation had improved Belinda couldn't bear to part with him. And then there is Millie, a chocolate-coloured border collie completely obsessed with fetching. She has worn herself out racing after frisbees, sticks and even pieces of grass thrown by excited children at the Mother's Day event, and has a sore paw.

Before Millie, there was a general rule that dogs weren't allowed in the house, let alone on the couch, and definitely not on the bed. Now Michelle and Berl sleep in a king-sized bed, purchased so there is room enough for the border collie, who has completely won their hearts. 'This house is a dog's life,' Michelle says, smiling as she takes in the prostrate canine forms.

Most weekends, Michelle and Berl will join them on the couch for a much-needed 'midday chill'. Berl once calculated that on average they each work the equivalent of 67 hours a week every week of the year. Although she has tried to cut back, Michelle is still putting in long days at the PCYC. She dreams of having more spare time to paint in the artist's studio set up in the backyard, and maybe even having her own exhibition one day, but she dreads the idea of retiring. In the police service, it's compulsory at the age of 60.

Belinda works long hours too, especially during the season, which starts in January with ground preparation and ends in early November, when the last crops are picked. There has been far less pressure since they stepped back from Halloween pumpkins. Belinda has even leased out most of the land and is trying to take each day as it comes, while she explores options for the next big thing.

Worried that Berl never has a proper holiday, Michelle bought her a tinnie as a surprise gift a few years ago in the hope it would encourage her to take more days off and go fishing—Belinda's favourite pastime. Stevi often joins her, and occasionally so does Odge, who taught her how to fish when she was much younger. Even though they are best mates, there is little conversation once they find a good spot. 'You don't need to talk,' Belinda says firmly. Michelle struggles with this concept and sitting still for any length of time, so she usually stays behind.

Nothing would make their friends happier than Belinda and Michelle taking advantage of changed laws in Australia that allow same-sex couples to marry. It's not widely known but they do have plans to tie the knot one day, they just haven't set a date yet. 'The ending hasn't been written!' says Michelle.

6

Tears in the Heart

NANCY WITHERS, SANDFORD, VICTORIA

On the June long weekend in Victoria, Casterton is the only place to be for kelpie lovers. The Australian Kelpie Muster is the largest gathering of kelpies and their aficionados in the world. Thousands of people pack the small town every year to celebrate the history and skills of this iconic working-dog breed, which holds a special place in Aussie hearts.

The program usually begins with a parade featuring a wide array of community floats and proud owners walking their dogs. The kelpies come in all shapes and sizes, from slim-hipped athletes who work for their kibble, to grey-muzzled retirees with portly physiques and family pets strapped into fancy coats that ward off the winter chill.

The real fun begins immediately after the parade, with the first event in a triathlon designed to show off traits for which the breed is highly prized. A 50-metre dash finds the fastest dogs. A high jump requires dogs to leap up and over a solid timber wall, demonstrating agility and courage. And then a hill climb tests stamina, discipline and independence as each dog makes its own way up a steep incline at the back of the town to find its owner. The competitor that performs the best overall is named King of the Kelpies, a fiercely contested honour carried with pride.

Mingling among the crowd most years is a petite woman, elegantly dressed, with short dark hair and striking eyes, who has devoted much of her life to preserving the rich genetic heritage of the kelpie, and trying to work out what makes these dogs tick. Nancy Withers is one of Australia's most highly regarded kelpie breeders and trainers. After almost 50 years of running her own stud and working with thousands of dogs, she remains passionate, curious, fascinated and awestruck in equal measure by these super athletes of the dog world, and their innate understanding of livestock.

Her path to unearthing their potential has taken Nancy from the excitement of training her first dog at the age of nine and working as a veterinary nurse in a large country town, to a cattle and sheep station on the salt-pan shores of Lake Alexandrina in South Australia and the verdant hills and gullies of Victoria's Western District. Along the way loss and illness have taught her to make the most of every day, while the generosity of strangers and an aristocratic black and tan kelpie called Bullenbong Mate, shaped her appreciation of dogs and their true place in the hierarchy of life.

Nancy's family roots lie deep in the rich volcanic soils of Mount Gambier's densely settled farming districts in the Limestone Coast region of southern South Australia. Much like the kelpie, her father's family were mostly of Scottish descent. Her great-grandfather, William Mitchell, left Perthshire as a young man in the late 1850s to join a brother who had taken up land just west of the town. Mount Gambier had only been founded a few years before but cultivated ground producing cereal, potatoes and livestock was rapidly replacing the districts' tall-timbered bush. The Mitchells played their part, helping to found the agricultural society and an agricultural bureau so that farmers might develop the skills and knowledge they needed to be successful in the new world.

Nancy's great-grandmother, Jane Telford, took a more circuitous route. She was born in a tent soon after her parents arrived in Adelaide from the Scottish border country. Her father found employment as a shepherd, then went dairying in the Strathalbyn district on land owned by one of the town's Scottish founders. Then in the late 1850s, the expanded Telford family packed their belongings into a bullock wagon and made the arduous journey down along the Coorong, with a mob of cattle. They established their own farm a few kilometres east of Mount Gambier at Glenburnie, where young Jane milked cows every day and made butter for sale.

Jane was 21 years old when she married William Mitchell

in 1861. According to family lore, she refused to walk down the aisle with him until he had at least a thousand pounds to his name, so he returned to Scotland and persuaded some aunts to lend him the money. William and Jane farmed in various locations before eventually fetching up at The Caves, a property on the eastern outskirts of the town.

Their son and Nancy's grandfather, another William, followed in the family footsteps and became a farmer too. He and his wife, Evelyn, were living on a property at Glencoe when Nancy's father, Frank, was born in 1912. The fourth of eight surviving children, he grew up to be an excellent sportsman and a capable student, turning down a scholarship to study pharmacy after he completed high school, because he was needed at home.

Frank loved his grandmother Jane, who lived with his family in her old age, but Nancy remembers his parents as being quite formal. As a child, she struggled to reconcile this rigidity with the great sense of humour shared by Frank and his siblings. 'Together as a family they were just so funny, and yet Grandpa and Grandma were always stiff. He always wore a stiff collar and these little round glasses, and Mum always said that Evelyn was a very hard woman,' Nancy says.

On her mother's side, Nancy's family were of mostly German descent. The Vorwerks were among the thousands of German migrants who came to South Australia in the early years of the colony to escape religious persecution. Some of the family settled near Mount Gambier in the 1850s, buying land south-east of the town at Square Mile.

Nancy's grandfather, Alec Vorwerk, married into another German family—the Eys. His wife, Edith, moved to Mount

Gambier from Adelaide as a child, with her father finding work in a government nursery that raised thousands of trees for forests and farms. She and Alec wed in the winter of 1912, when they were both in their early twenties, and had five children, including Nancy's mother Bette, who was the middle child.

Bette was only six and the youngest no more than two years old when tragedy struck. At the age of 36, Edith died of tuberculosis. She became ill after nursing a man no-one else was prepared to help for fear of catching the highly infectious disease. Unable to cope on his own, Alec sent Bette to live with her Grandmother Ey, an eccentric Irishwoman who resided in Melbourne. 'That had a huge effect on my life I think, because my mum valued family so much and was a very compassionate person. She really felt the loss of her mother, and being separated from her family and going to live somewhere else. Family was precious to Mum,' says Nancy.

It was six years before Bette returned to live with her father and his second wife, Nell, after they became concerned that she wasn't being looked after properly.

Bette grew up to be a striking blonde, with a wide smile and long dark eyebrows arched over green-brown eyes, but it was her older sister, Gladys, who first caught Frank Mitchell's attention. Then he started noticing Bette. Nell made her work hard to earn her keep, and she always seemed to be doing chores whenever Frank came to take Gladys out, so he felt sorry for her. On her part, Bette first became aware of Frank when she heard all the girls talking about him one night at a dance. 'Well, I don't think much of him,' she commented dismissively, after he was pointed out to her.

More than 180 centimetres (just over 6 feet) tall, good looking, with thick dark hair and broad shoulders, Frank was quite a catch. His parents were successful farmers, well known in the community, and he was something of a local sporting hero in both football and cricket. Frank won the Blue Lake Football Association's Hirth medal as the season's best and fairest player on two occasions and in 1936 led the Yahl team to its first premiership win.

His track record in cricket was more controversial. A noted fast bowler and consistent batsman, he scored more than a thousand runs with the bat in 75 innings, one season ranking as the association's top batsman and hitting 47 runs off one over. But in 1934 Frank found himself at the centre of a major to-do, when his intimidatory bowling was compared with that of infamous British cricketer Harold Larwood. The English bowler had notoriously used the 'bodyline' style of bowling against Australia during a recent test series. Believing Frank's style was similar, the umpire no-balled him seven times in one match, applying for the first time a new rule introduced to prevent bodyline tactics being used.

The controversy reached its climax a few months later in the final. Frank took three wickets for 10 runs, clean bowling two of his victims. After being rested, he was brought in again at a critical stage of the match and received a warning from the umpire in his first over. In his second over, the umpire barred him from further bowling after eight no-balls. At the time, Frank was hurling down deliveries to his brother, Bob, who didn't know what the fuss was about, explaining that Frank's bowling 'had no terrors for him'. The controversy made headlines in Melbourne and Adelaide newspapers and upset Frank greatly.

'Dad never forgave that umpire. He was one of the best and fairest sportsmen in the South East at the time, and it bothered him greatly the rest of his life,' says Nancy.

After getting to know Bette a little better, Frank decided he really liked her and, some time later, he asked her out. They married at St Andrew's Presbyterian Church in Mount Gambier in March 1939, only months before the start of the Second World War. Nancy is convinced her petite mother wore very high heels for the ceremony, or was posed standing on something. In the wedding photos, she doesn't appear that much shorter than Frank, but she was only about the same height as Nancy, who is 157 centimetres (5 foot, 2 inches) tall.

The newlyweds moved onto a farm at Square Mile, where they ran some dairy cows that Bette helped milk twice a day. Farming was a reserved occupation during the war, so Frank didn't enlist. With labour in short supply because so many men were away fighting, he held down several jobs, working at a local abattoir and shearing sheep for other farmers, as well as helping to manage The Caves.

Frank and Bette had given up on the idea of having children, when Bette fell pregnant with Nancy. Their daughter was born in November 1950, during the middle of a ferocious thunderstorm. 'Which might explain why I like storms. I just love it when the weather is rough!' Nancy adds, with a mischievous smile.

Nell and Alec were delighted with their new grandchild, but the same could not be said for Frank's parents. 'Never mind Bette, better luck next time,' they told her, after learning the baby was a girl. It was shearing time at The Caves, and when Bette arrived home from hospital, she found a note from Evelyn to say they had gone to their beach house south of Mount Gambier, at Port MacDonnell, for a holiday. The new mother was left on her own to look after the shearers, as well as the baby.

In 1954, Frank and Bette had a second child, Craig, and then, in 1958, Allan was born. By this stage, they were living in town, while Frank was still managing The Caves for his parents as well as leasing several blocks of land around Mount Gambier with his father-in-law, to run sheep and grow potatoes. One of Nancy's earliest memories is riding out to one of the farms in Frank's small truck. The vehicle also brought about the first interaction with a dog that she can recall. Nancy was about four when Frank lifted her up and placed her on the tray, alongside his border collie, Mac. The dog took umbrage that the child was sitting in his usual spot, directly behind his master, and nipped her on the arm to move her over. 'Mac was senior and I was the usurper!' says Nancy, excusing the dog's behaviour.

During her childhood, Nancy also spent a great deal of time with Nell and her Grandpa Vorwerk. She adored Alec, who was an important influence on her life until he died in 1965, at the age of 77. 'He was just cuddly, and warm and loving,' Nancy says. 'I don't think he ever got over the loss of his first wife. I looked like her so he had quite a soft spot for me, and he took me everywhere.' Some of her favourite moments were spent in Alec's workshop. He was quite an inventor, applying his practical skills to designing and improving farm machinery. He also loved carpentry and making furniture, which became a preferred pastime in his old age, and he patiently showed Nancy how to mitre joints and assemble them.

Nancy's love for training dogs began in heartbreaking circumstances. When he was only six years old, her younger brother Craig was diagnosed with leukaemia. The family had just come back from a holiday at Port MacDonnell. Time spent at the beach usually revived everyone's energies but Craig seemed tired and lethargic so Bette took him to the doctor. Blood tests revealed the fatal illness, with the doctor predicting he might have only twelve months or so to live. 'You can imagine what that did to Mum. It was a real upheaval in our lives,' says Nancy.

Determined to do what she could for her son, Bette took Craig to Adelaide regularly for extended visits so he could receive blood transfusions, leaving Nancy behind with her father and Allan, who was only a toddler. Within months of the diagnosis, the Mitchells sold their house to cover the medical costs, and moved into a house behind a continental deli shop in a northern part of the town. The business provided a source of income that enabled Frank to keep an eye on the children while his wife was away. Many of the shop's customers were Italian migrants, who came to Australia after the war. Having learnt Latin at school, Frank soon picked up their language and made friends in the community who were very supportive. 'One of the fondest memories I have of that time, and there isn't a lot of them, was going to family celebrations and singing songs in Italian. They were beautiful families,' Nancy says.

She was not so fond of the Latvian housekeeper her parents engaged to help Frank when Bette was away. Nancy took a strong dislike to the woman and missed her mother enormously. As a form of compensation her parents bought her a dog. 'I was nine, and he was my first dog,' Nancy explains. 'He was a border collie cross, black with a little bit of white on him, and I called him Peter.'

Nancy loved the dog but in truth he was little comfort when Craig died in October 1960, twenty days short of his seventh birthday. 'Mum absolutely insisted that he die at home, and he did, in her bed, being held in her arms,' she says, becoming visibly upset at the memory.

Nancy was almost ten at the time, and Allan, who she calls Mitch, had just turned two. Considered too young to attend the funeral, both children were sent to stay with the Vorwerks for a few weeks. 'My father didn't discuss with me anything about those years and how he felt, never ever, but Craig's death really affected my life, because after that Mum buried herself in work. With her background, with her mother dying and her love of family, to have that happen it just blew the bottom out of her world. She was still a wonderful mother, she just had to bury the grief somehow. Allan was little so he needed attention but, basically, I was left to myself and that was fine.'

The shop was sold and Bette found employment managing a clothing store. A talented seamstress who had been earning money sewing since before she was married, Bette also kept making and altering garments for private customers. She specialised in wedding dresses and sometimes designed garments too, instead of simply following a pattern. Nancy often helped her mother bead the gowns and match the lace.

A competent cook, Frank didn't mind preparing meals while Bette worked, and it wasn't unusual for him to do the washing and vacuuming. In his spare time, he remained heavily involved in sport, coaching teams and individual athletes, or he could be found reading. Frank passed his love of literature on to Nancy, whose house is full of books. 'He often used to quote Shakespeare and talk about Greek mythology, and he would really make you analyse everything. He taught me to be analytical from when I was very young. He really was a very interesting man . . . He said to me once, "In life, don't go out to beat other people, just try to do better than your best. Beat yourself, that's all you have to do." He had the most amazing way of teaching us, as children.'

Thinking more about her father, who was clearly a strong influence in her life, Nancy admits: 'Mum used to get a bit frustrated with him, because he was a very passive person. When he was 93, he came to live in Casterton and I got to know him a lot better. I realised why Mum thought he was frustrating because he wouldn't discuss things. With my dad, and I'm a bit the same, by the time something was discussed, he had already thought it all through from every angle and made his decision based on what he believed was fairest . . . He didn't ever take a dislike to people, but when someone disappointed him he dealt with it by quietly withdrawing his services. He would never say anything—he would just quietly not be available.'

As she grew older, Nancy was often given charge of her little brother, who gave her some heart-stopping moments. One day, when Mitch was about five, she was instructed to take him to the Odeon theatre in the centre of town to see a matinee movie. The theatre was packed with parents and children, and she lost him in the crowd when everyone was streaming out into the street after the session ended. 'I could hear this wailing noise in the distance and I recognised the cry. I fought my way through the crowd, and there he was, the little bugger! And when he was four, I can remember being in [the main street] when he flopped down and started screaming. You know that stage when kids throw tantrums? I didn't know how to deal with it, and I'm trying to drag him along by the arm.'

Nancy and Mitch are close now, and often joke about these moments, but she thinks he was spoilt after Craig died. 'He was our mother's darling and that is his biggest problem,' she says, laughing. 'But he was very precious to Mum, you see. Her parents-in-law had made her feel she had to have sons, and then she lost the first one.'

By contrast, Bette didn't seem to worry too much about her daughter. Nancy remembers having an accident on a new swing installed at a park near their house. When she and Mitch got back home, their mother was fitting a customer for a new dress so she told Nancy to go to her room while she finished, not realising how serious the injury was. 'After a while Mum must have thought, "She did look a bit pale," and she came in. I was still sitting on the bed, and she said, "Oh my goodness, you've broken your arm." I had a yellow jumper on, and I could see the arm had this great kink in it.'

During these years, Nancy spent a lot of time with extended family. Aside from her Grandpa Vorwerk, she enjoyed visiting her uncle Frank Spehr, and his wife, Rae, who were dairy farmers a few kilometres south of Mount Gambier. Frank had a black collie-kelpie cross called Jet, and a long-haired black and tan dog called Guy, who were trained to bring the cows in for milking by themselves.

Then there was Nell's sister, Alice, and her husband, Jack Atkinson, who was extremely successful competing in sheep-dog trials with his border collies, winning a Victorian state championship and numerous other competitions across the region. Jack kept his dogs in large runs made out of pine offcuts sourced from a local timber mill, and trained them using sheep held on a block next to his house. In a formative experience for the young girl, he introduced Nancy to dog trials when she was only five or six, and often took her to competitions. 'I used to toddle around behind Uncle Jack and lead the dogs at the trials, and he would explain to me what they were doing. I really loved a dog called King Nero, a tri-coloured border collie, and then there was Mullana Lady, Easter Lady and Bright Speck,' she lists, demonstrating an almost photographic memory for the hundreds of dogs she has known in her lifetime, and their bloodlines.

Nancy may have been obsessed with animals—dogs and ponies in particular—but she was also a bright and capable scholar. She began her education at the Mount Gambier Primary School in Wehl Street, where her teacher was a tall spinster, who the children called Crab-apple Creed. After it opened in 1956, she transferred to the new Reidy Park Primary School and then she moved again to another new school closer to home when it opened during a period of booming growth in the town. 'I was in Year 7 and I had a teacher called Mr Hammond. I spent a lot of time cleaning fish tanks, which I loved,' she says.

Nancy missed out on being dux of the school by just two marks. At high school, she was placed in the advanced class and focused mainly on science subjects. On the insistence of her father, she also took Latin. For much of the time, her class was taught by the headmaster and headmistress—not because the class was smart but because the students were badly behaved. 'We were dreadful. One woman had a nervous breakdown trying to teach us,' Nancy admits sheepishly.

Nancy usually sat at the back, not because she wanted to make mischief but so she could unobtrusively spend more time reading her favourite books, leading to dire predictions from one particular teacher. 'I was reading a book during chemistry, under the desk up in the back corner, and he asked me a question and I had no idea what he was talking about. And he said to me, "If you pass chemistry, I'll eat my hat." I passed chemistry quite well, and I was so upset because he had already left the school.'

During her teenage years, when she wasn't doing homework, or playing netball and tennis, Nancy spent as much time as possible riding horses. They were borrowed from friends and family, because her parents refused to buy her a pony of her own, anxious that she not become a 'horsey woman'. Two horses from that time stand out in Nancy's memory—a skewbald mare called Kim, and a handsome galloway named Socks. A strongly built pony with a lovely temperament, Kim would bust herself to keep up with larger horses. 'She wasn't pretty but you could do anything with her. I rode her bareback over jumps, and she was totally trustworthy.' Years later, when Nancy was married with children of her own, the owners gave her the mare, who lived until her thirties, which is a great age for a horse.

Socks was also a skewbald, mostly bay in colour with a black mane and tail. 'He was the prettiest horse and I just adored him. I must have ridden him on and off for eighteen months or so and then the people who owned him decided they wanted him back, so they took him to where they lived out of Mount Gambier. Two days later I heard a noise outside and I got up, and Socks had come back to me. He was on the front lawn.'

Realising Nancy had a gift with horses and there was no point trying to discourage it, Frank bought his daughter a thoroughbred mare called Gypsy. The horse had been prepared for racing, but before she was ready to compete her owner-trainer was hit by a car and had his legs broken. Gypsy had been his wife's pet and she didn't want to part with the horse but Frank must have persuaded her. When it came time to ride the mare home, he couldn't do up the surcingle on her saddle because she was so fat.

The Mitchells were still living in town, so Gypsy was agisted in a paddock within the city limits. 'Of course, I spent every spare minute on that horse,' confesses Nancy, who was fifteen when the mare entered her life. 'Gypsy was proud, she was fast and she knew it, and she was kind but she was flighty and very powerful. I don't know how I'm sitting here, some of the things we did. My brother teases me to this day. I can remember one incident when we were riding along a road. He was on Kim and he came up alongside me, and got just that little bit in front. Gypsy wouldn't be beaten—as soon as something poked its nose in front of her, she would be reefing and have to go. She would gallop for a mile and she was that fast you could hardly see where you were going.'

A couple of lads involved in the pony club knew the horse had this tendency and would deliberately exploit it for a bit of fun. Small and slight even compared to a jockey, Nancy found herself tearing along a back road during an Easter pony-club camp one year after Richard Wilson and Peter O'Connor shot past her on their thoroughbreds. 'They were wild riders those fellas, great horsemen, and they used to take delight in causing Gypsy to bolt. I'd be gone for all money but I was never frightened. I didn't mind going fast.'

That incident ended without horse or rider coming to harm, but that wasn't the case when Nancy was riding along a different road one day, leading another horse and accompanied by a friend riding Kim. They were cantering along when Gypsy accidentally put her front leg down an old post hole, left unfilled after a road-sign had been removed. 'I went sailing off and crashed into the ground, and the other horse came down around me. We both fell off in this big tangle.' Nancy was extremely relieved to see the horses get up and trot away, but she ended up with five cracked ribs. 'I didn't realise it, except I was terribly sore on my side. We had to catch the horses and someone stopped to help us, then we rode home.'

When Nancy was still at school tragedy struck her extended family. Bette's brothers owned a shack at Pelican Point, about 40 kilometres west of Mount Gambier, where they loved to go fishing just offshore in a small wooden boat. In August 1965, her youngest brother, Ted, went out with his two sons—Brian, who was seven, and five-year-old Lex. Something happened and the dinghy flipped over. Ted wasn't wearing a life jacket because it had been stolen, but the boys had theirs on and he managed to lift them out of the water and sit them on the upturned boat. Ted was pushing it towards the shore and had reached water shallow enough to stand up in, when another wave hit and he was knocked out.

The current kept carrying the boat closer to shore, so Brian grabbed his brother and dragged him off the hull, intending to pull him to safety. Shortly afterwards, a man fishing nearby heard a boy crying out for help. He found Brian just as he was being washed ashore. The boy was safe but there was no sign of Lex or Ted. After a frantic search, Ted's wife, Rhonda, found Lex floating facedown in the water. He was rushed to hospital in Mount Gambier but it was too late; he was pronounced dead later that afternoon. Alerted to the accident, Frank and Bette rushed down to Pelican Point, leaving Nancy to stay by the phone. With darkness closing in, police were about to call off the search for the night, when Frank found Ted. He was dead too.

A month later, Nancy's beloved grandfather, Alec, died after having a heart attack at the age of 78. 'I think it was too much, losing Ted like that,' Nancy says with great sorrow.

These losses, and the death of her brother Craig, forged in Nancy a determination to make the most of life. 'There is not enough of life to waste. You are lucky if you are granted the gift of life and you should do the best with it that you can, the most you can,' she says with fervour.

Despite being a bright scholar, Nancy decided to leave school after her intermediate year and become a nurse. She had the potential to be a scientist, but in those days girls weren't really encouraged to think beyond traditional careers as teachers or nurses. Besides, her mind was made up after witnessing the care and concern nursing staff had shown her mother and Craig when he was ill, and she liked the idea of helping people.

Nancy had to wait until August 1967 to begin her training, because student nurses couldn't enrol until they were sixteen years and nine months old. Hopeful that it might encourage her daughter to take an interest in more feminine pursuits, Bette used the intervening period to send Nancy to the famous June Dally-Watkins finishing school in Sydney, where young women were taught etiquette and deportment. Nola Warren, a cousin of Bette's, worked for June, who had also established Australia's first model agency. Nancy remembers meeting Nola for the first time a few years before going to Sydney. The beautiful brunette was something of a celebrity, having been a top model and one-time actress, after securing a part in the Australian film *White Death*, starring famous American author, Zane Grey, when she was a teenager.

Starting on a salary of $14 per week, Nancy was one of twelve student nurses in her intake at the Mount Gambier Hospital, which sat high on a hill overlooking the city. Nancy only knew one of the other girls at the beginning, but that soon changed. Aside from working together, they were all expected to stay in the nurses' quarters next door, even if their homes were nearby.

Nancy enjoyed the training, and didn't really mind Molly Ogden, who was matron in charge at the hospital for more than twenty years. However, she often found herself in trouble with the senior nursing sisters. According to the rules, nurses must never run, except in the case of fire or a patient haemorrhaging.

This decree applied even when moving from the hospital to the nurses' quarters. Physically active and always in a hurry, Nancy often forgot and broke into a trot. Many a time one of the senior sisters would spot her and query: 'Nurse Mitchell, fire or haemorrhage?'

Despite the strict rules, Nancy soon realised some of the senior nurses had a sense of humour behind their stern exteriors. One night she went to a dance at the Barn Palais, a popular night spot a few kilometres out of town, and met an attractive young man who was visiting the area to compete in a surf-lifesaving carnival. He wanted to see her again, so she told him to drop by when her late shift was due to finish the next night. Men were definitely not allowed to visit the nurses in their rooms, and there was a strict curfew that locked nurses out if they returned too late on their nights off, but Nancy thought it would be okay if they talked for a while.

She had just come off duty and was walking towards the nurses' quarters with another nurse and one of her supervisors, Sister Blight, when the young man leapt off the roof of a garage they were passing and landed in front of them. 'Here's this tall blond hunk and he's smiling at me. And Sister Blight says, "Nurse Mitchell, do you know this young man?" I could tell they were trying not to laugh.'

Nancy loved nursing, although she found it emotionally draining at times. Aside from attending lectures, every student nurse was expected to spend two months on each ward as part of their training. Nancy's first assignment was the men's surgical ward, followed by women's medical. Then she found herself stuck on the men's medical ward for six months, including a stint on night shift. Spending so much time there took its toll because many of the patients had terminal illnesses, such as cancer, and died.

Nancy recalls one particular incident when a doctor and nursing sister told a patient that he had terminal cancer, and not long to live. 'Then they left him to it. I was the next person to walk into the room, and he just put his arms around me and he cried and cried and cried. I was seventeen. I was used to dealing with death and grief but it was a tough gig and we had no counselling whatsoever.'

To fend off the tears, the nurses developed a slightly macabre sense of humour. Nancy remembers an elderly man being admitted to a four-bed ward just as visiting hours were about to start and people were lining up in the foyer of the floor where she was working. After he was settled in his bed, the patient asked her to open the window and let in some fresh air. When she turned around from this task, he was dead. 'He had died, just like that, and I thought, "What do I do now?" I went and told the sisters, and they decided the best thing we could do was put an oxygen mask on him and a little theatre cap. Then we whizzed him straight down through the visitors and into a single bedroom. By the time we got to the room we were all giggling.'

Nancy's nursing career came to an abrupt end not long after she passed her first-year exams. She was riding a recently gelded horse when it started to buck. The saddle slipped over the horse's withers as he spun around, with his head facing down a slope. Nancy was pitched onto a metal road, somehow bending her left knee over the heel of her right foot as she landed. The ligaments in her knee were damaged and she couldn't walk properly. The medical superintendent at the hospital diagnosed her with synovitis, ordering her to take two weeks off. She still couldn't climb steps properly by the end of that period, which was a considerable problem in a multistorey hospital where the nurses often had to take the stairs because the lifts were busy.

Deciding to take a full year off nursing, Nancy found work as a nanny for the McLachlan family, who own Nangwarry station, about 40 kilometres north of Mount Gambier. She had care of Ian and Janet McLachlan's daughter, Edwina, who was about four, and their eighteen-month-old son, Dugald. 'It was one of the happiest years of my life,' Nancy says. 'It was a lovely homestead with big red gums, and the family were very inclusive.'

And there were the horses! Most days, Nancy had the opportunity to assist Ian's groom—a South African woman called Judith. Despite her damaged knee, Nancy was able to ride, providing there was help at hand to get into the saddle. In the early dawn, she would make her way to the stables and help exercise the string of four or five polo ponies Ian kept. She and Judith would go on long rides through the pine plantations near the station, revelling in the landscape and the soft morning light as the sun rose.

In the evening, Nancy usually joined the McLachlans for dinner, along with the rest of the staff. Aside from Judith, who did some of the cooking, there were two stockmen and two students from Marcus Oldham agricultural college in Victoria, working at the station for a year as part of their practical studies. One of those students was Tim Withers. They didn't really socialise because while Nancy lived on the station during the week when she was working, she usually spent her days off at home, with Bette driving up to collect her. Then, about two weeks before Tim was due to leave, he asked her out on a date. They continued to keep in touch and go out together even after Tim returned to college at Geelong and Nancy finished up at the station.

With her knee now better, Nancy had every intention of going back to nursing, but she was offered a full-time position as a veterinary nurse with Dr Geoff Manefield, who had an expanding practice with surgeries in Mount Gambier and Penola. Nancy loved animals and nursing so the move made perfect sense. Dr Manefield sent her to the Institute of Medical and Veterinary Science laboratory in Mount Gambier to learn how to carry out basic tests. As his theatre sister she assisted him with post-mortems and operations, and she was made responsible for keeping the ledger and paying accounts. 'Geoff was a very clever man—they still use some of the procedures he pioneered, to this day. I worked for him for two years and it was a very interesting and happy time; and I could walk home from work, and go to Aunty Nell's just across the back for lunch.'

When she was almost 21, Nancy left the practice to get married. After maintaining a long-distance relationship, Tim had proposed and invited her to visit Nalpa station, where they would make their home. Situated on the western shores of Lake Alexandrina, about 30 kilometres south of Murray Bridge, Nalpa had been in the family for three generations. The 5260-hectare property was well known for its shorthorn beef cattle, which ranked among the best in Australia. Tim's grandfather, Alfred Withers, bought the station in 1922 and a few years later established a stud based on champion blood-lines, which he further improved by importing a number of sires from Scotland. By the time Nancy visited, the property was carrying around a thousand head of cattle and about three thousand sheep.

The country was very different from what Nancy was used to. Despite being not that far from a large regional town, it also felt isolated. On her first visit, she travelled up by bus, along the Coorong, and got off past Meningie, at a turnoff leading to Wellington, where a ferry crosses the River Murray. Tim was supposed to meet her, but there was no sign of him. It was a hot, windy December day and there was no shade so Nancy sat on her suitcase at the side of the road and waited, one hand clamped to her large straw hat to stop it blowing away.

Tim pulled up in a cloud of dust 30 minutes later, having been held up on the other side of the river, waiting in the ferry queue. They waited another hour to get back on the ferry, and then drove towards the homestead. Its location was marked on the horizon by a distant clump of trees—the only trees Nancy could see for miles. 'I looked around at the sparsely covered plain, at salt lakes and samphire, at the big lake in the distance to the south and the haze of the Mount Lofty Ranges in the west and tried desperately to think of something complimentary to say about this land so steeped in his family history,' Nancy later wrote.

Determined to make the most of the visit, she joined Tim on his daily rounds checking water troughs—a critical task in summer. Taking an old two-seater jeep with half doors, the young couple crossed a main road and discovered a mob of cows with calves at foot. They had pushed through a boundary fence and were grazing in the adjacent reserve where there was no access to water. Most of the mob obligingly walked back through a gate into the paddock, but one cow was so crazed with thirst she jumped the fence, bolted across the road and then jumped another fence into the neighbouring property. Leaving Nancy to make sure the rest of the mob stayed put, Tim raced off after her.

When he tried to guide the cow back, the enraged animal charged the jeep, slamming her horns into the side. He backed off, giving her time to calm down, then drove to collect Nancy before the cow spotted her. 'I had barely reached the seat when she charged with full fury and hit the half door, denting it badly, only inches from my left arm. As she raised her head, I got a full view of her blazing eyes and could feel her breath. By then, I was on Tim's lap in the driver's seat . . . Later I pointed out to him that some men will go to no end of trouble to get a young lady into their lap!'

When Nancy moved to Nalpa in July 1971, she became part of an extended station family. Tim's parents lived in the main homestead. His older brother, Philip, lived on Nalpa too, with his wife, Mary, and their two children. A succession of workmen and their families also came and went over the years.

Nancy and Tim's first home was a very basic house built some years before for the station overseer. Made out of concrete bricks and set on a low sandhill, it became incredibly hot in summer and then cold in winter. 'For the first summer, at night I lay on the lino in the hall trying to get cool. We had no air-conditioning, no fans, no anything. Then we bought this fan and it cost $34. It took us three months to save up for it.'

Although she was used to cooking on wood stoves, Nancy installed an electric oven, which helped to keep the house cooler in January and February when she took advantage of seasonal fruit to make jams and preserves. Otherwise, she spent as much time as she could out on the property, where the Withers family were happy for her to participate in the farm work. 'In those days a lot of women were not allowed out on the big properties, but I must say they were terrific. I would have gone crazy if I'd been stuck in the homestead,' Nancy admits.

She'd had very little to do with cattle before, but she came to appreciate them and was fascinated by the bloodlines at Nalpa, stretching back to the 1920s. In 1973, the Withers also established one of the first Simmental herds in Australia, with Tim learning how to AI cattle so they could produce cross-breds for the beef market. Nancy soaked up his considerable knowledge of cattle and helped train all of Nalpa's stud bulls so they could be led calmly into an auction ring or compete at the Royal Adelaide Show. Despite this training, they didn't always behave, with Nancy making more than a few rapid laps at the Wayville showground over the years, dragged by animals that weighed as much as a tonne.

As her first experience on Nalpa demonstrated, things didn't always go smoothly. One day, she was driving the jeep to muster cattle while Tim worked from a motorbike. Attempting to turn the vehicle and head off some breakaways, she discovered the steering wasn't working. The vehicle kept going straight ahead and the cattle got away. 'What the hell do you think you're doing over here?' an irate Tim asked after roaring up alongside. Nancy descended slowly from the vehicle and with great dignity asked him to take a turn driving the jeep. He discovered the steering had a broken kingpin. Adopting the subdued tone he often used instead of apologising, Tim politely asked her to go after the cattle on foot. Still keen to be helpful, she agreed. 'Two kilometres later, puffing away, I briefly wondered at my gullibility and decided that those huge blue skies and all that fresh air had gone to my head!'

Another mustering exercise where the Murray enters Lake Alexandrina taught Nancy the dangers that come with working cattle in tough terrain. Tim didn't like thoroughbreds, even though his grandfather had been a racing man and kept them on the station, but Nancy had insisted on bringing her beloved Gypsy to Nalpa. 'I won't marry you unless I can take my horse,' she'd told him when he proposed.

This particular day Nancy rode out alongside Tim, who was mounted on Gypsy's daughter, Honey. The aim was to bring in cattle grazing on an area the family called Island Point. Half bog and half reed beds, the narrow neck of land joins the mainland to Pomanda Island, a triangular island about 34 hectares in size. The area offered good summer grazing but it was liberally sprinkled with boxthorns and inaccessible by vehicle.

Riding together, Nancy and Tim made their way along a narrow track, dodging the deeper parts of the bog and its treacherous mud. After crossing a creek that flowed belly deep around the horses, they came to an open stretch of land where they expected to find some cattle. There were none in sight so they walked on, splashing through another creek and scrambling out onto the harder ground of the island itself. The two riders separated to work different areas and push any cattle they found back towards the mainland, but Nancy soon realised Pomanda was no place for Gypsy. Almost impenetrable in places, the boxthorns cut into her thin hide as she darted and baulked in turn. Eventually horse and rider emerged into a more open area, where they found half a dozen shorthorn cows and a big roan bull.

Nancy had them heading in the right direction when a large white heifer shot out of the bushes. Gypsy went to give chase but Nancy managed to settle her, and turned towards home. Waiting on the creek bank for Tim, she looked down to see a deadly black snake slither between her horse's legs and into some reeds. Relieved that the horse hadn't noticed but feeling even edgier after the experience, Nancy decided to cross the creek and wait on the other side. 'Then all hell broke loose. The white heifer shot out of the reeds on my right and charged past us into the lake. Gypsy shied violently and bounded up and down, reefing at the bit as I fought to calm her.'

Nancy knew she had lost the battle when the horse lurched back into the creek and she was forced to slacken the reins so the mare could regain her balance. Gypsy took off, heading for the bog, where she made a sideways leap and went down. Nancy's ears drummed as the muddy water swirled over her head. She surfaced with her legs under the struggling horse, then Gypsy was up and off.

Nancy learnt years later that she had broken her collarbone, most likely during this fall. Conscious only of a painful right shoulder, she climbed out of the bog and set off to find

Tim. He was back near the creek, only marginally drier after his own horse had been spooked by the same heifer. Smiling at her wryly, he reached out and plucked waterweed from her drenched hair.

Nancy eventually located an exhausted Gypsy at the next fence, stiff and sore like her rider. Together, they slowly made the three-kilometre journey home. The horse remained part of Nancy's life until 1974, when she died six weeks after delivering a foal. Before it came, Gypsy was being kept at a stud in the Barossa Valley where she ate Salvation Jane. Also known as Paterson's Curse, the weed is toxic to horses, causing liver damage and photosensitivity.

Gypsy's death in such terrible circumstances upset Nancy greatly, compounded by the timing. Sixteen days before Gypsy gave birth, Nancy and Tim welcomed their first child into the world—their son Simon. Nancy now had a young foal to care for as well as her own baby and a valuable kelpie pup she had taken delivery of just a few months before. 'Someone told me goat's milk mixed with raw sugar was great for rearing foals, so I got a goat. So then I had a dog, a goat, a foal and a baby to look after,' she says.

In 1976, Nancy gave birth to another son, Jonathon. He arrived close to the expected due date after a false alarm when Nancy experienced Braxton Hicks contractions while cooking for shearers at a family property at Keith, with help from her mother. The shearing team had moved on to Nalpa when Tim came home on a Friday evening about two weeks later to report they were taking bets the baby would arrive before Monday.

His good humour evaporated an hour later when he developed a migraine so severe that he vomited repeatedly.

Nancy got up to check on her husband at around ten o'clock that night and her waters broke. Finding Tim kneeling over the toilet bowl, she soon realised there was no way he would be able to drive her to the hospital about an hour away, over bad roads. Blessing one of the advantages of living on the property with extended family, Nancy phoned her brother-in-law and within minutes they were on their way to the Murray Bridge hospital where an officious nurse met them in the corridor.

'Mrs Withers?' she checked.

'Yes,' Nancy replied.

'Mr Withers?' she asked, turning to Phil.

'Yes,' he replied automatically, not realising the assumption that had been made, until the nurse looked confused when Nancy ordered him from the room before she agreed to undress.

By six o'clock the next morning, the birth was imminent. Nancy asked the nurse to phone Tim and get him to come, then she was given a pethidine injection. 'As it turns out I can't have pethidine, and it absolutely flattened me. It knocked me out totally.' She came round from the drug before Jonathon was born, but his breathing was depressed.

Nancy blames this experience for her son being diagnosed with asthma as a child, and the sleeping problems he experienced for much of his childhood. 'He didn't sleep through the night until he was three years old. It was nothing for me to get up to him three or four times, and on a very bad night it might be seven times. Like so many men of his generation, Tim didn't believe it was his job to get up to the children, so it was left to me,' she says.

Completely exhausted, Nancy then had to drag herself out of bed early to prepare Tim's breakfast. 'I don't have an addictive personality, but I used to think if there was one thing that I could become addicted to, it's sleep.'

Life improved for Nancy in 1980, when the family moved to the South East where they bought an 810-hectare property between Kingston and Robe, which they named Pomanda, and hired some live-in help. Still a good friend, Chrissy got up to the children on alternate nights so Nancy could rest properly. Then came the morning when they realised that neither of them had been woken by a crying child. 'We ran in to see if Jonathon was still alive! He was three and it was the first time he had slept through the night.'

During those exhausting years, Nancy had a major health scare. Deciding she couldn't cope with any more children and unable to take the pill for medical reasons, in 1978 she went into hospital for a tubal ligation day procedure. She woke up feeling like a truck was parked on her chest. 'I couldn't speak. I was basically paralysed from the muscle relaxant they gave me for the operation. They had brought me round, thought I was okay and put me in the recovery ward. I didn't panic, but it was like this massive vice had hold of me and I couldn't tell anybody,' she explains.

Eventually the drug wore off, and Tim took Nancy home. Not that many weeks later, she experienced the worst pain of her life, like a hundred crocodiles were gnawing her stomach. After about half an hour, the pain wore off, leaving her feeling badly bruised around her middle. Nancy went to see her doctor who was baffled. The attacks kept returning, some worse than others, but she didn't live close enough to reach the doctor while they were in progress so he could get a better idea of what was happening.

'There was no sign anything was wrong except that I was losing weight. I went to I don't know how many different doctors and specialists over a period of eighteen months, and by the time I'd finished I was 6 stone 2 pounds (39 kilograms), and still getting up all night to a child, and looking after my husband and Simon. Most of the doctors treated me as if I had a mental health problem but I must admit Tim was very good. He never doubted for one minute that there was something wrong.'

Eventually, Nancy's doctors decided to remove her gall-bladder. When the surgeon could see nothing immediately wrong with it, he decided to remove her appendix too before bringing her round from the anaesthetic. Nancy was back in recovery when she stopped breathing. Medical staff managed to resuscitate her and inserted an intubation tube, but she continued to have trouble breathing. 'I wasn't in a very good way,' she says. 'Then six weeks later I had the worst attack of pain I'd ever had.'

It took two years before a doctor eventually diagnosed the problem. 'Your gut is paralysed; there are no gut movements,' he told her after she reached the Meningie hospital in time to be examined during an attack. He gave her an intravenous drug to help her gut relax and the pain reduced immensely.

Now living in the South East, Nancy went to see Kingston's resident doctor, who suggested that she keep a record of everything she ate immediately before an attack. It turned out that a major trigger was corned beef, because of the chemicals used in the preserving process. Bacon, ham and cordial also caused problems for the same reason. 'Once I cut all those things out of my diet, I was heaps better. I would still have moments when I felt weak, but I didn't have bad attacks.'

Then in December 1983, Nancy became seriously ill again, with Guillain-Barré syndrome, a rare and debilitating disorder in which the body's immune system attacks the nerves. The first notable symptoms came when she was driving home after visiting her parents in Mount Gambier. She dropped in to see a friend along the way, and was walking down his driveway when her forearms started aching. 'The next morning when I woke up and put my feet on the floor to get out of bed, it felt like ants were eating my feet. By halfway through the day, it was halfway up my calves.'

Her doctor thought it might be Ross River virus but the tests came back negative. About three weeks later, within a period of just 24 hours, the sensation spread to her thighs, hands and arms. When she rang her doctor again, he ordered her to go straight to the Mount Gambier Hospital, where she was admitted to a four-bed ward. Nancy didn't get a wink of sleep that night, and nor did the three older patients sharing the room. 'All I did was moan with the pain, and they couldn't give me any drugs because they didn't know what was wrong with me. It just went on all night, getting worse. By the next morning I had no feeling from my fingertips to my shoulder blades, and from my toes to my hips. And that's where it stopped, which was lucky—some people end up having paralysed breathing muscles.'

There is no known cure for Guillain-Barré syndrome, but having diagnosed the problem doctors at the hospital did what they could to ease the symptoms. It's not uncommon for sufferers to be unable to walk for months, or to experience lifelong effects such as fatigue and numbness. Determined this wasn't going to happen to her, Nancy refused to stay in bed, constantly working to get her feet moving again. As soon as she could, she returned to the farm, where Gypsy's grown-up foal, Breezy, became her legs. A workman caught and saddled the horse, and Nancy rode everywhere, continuing to do stockwork.

After a few months, feeling gradually returned to her limbs. Unfortunately, the painful sensation that ants were eating her nerve-endings also came back. It continued for almost another two years. 'I used to feel like screaming and crying some days, but I didn't do either. I had a young woman, Rae, who was a great help in the house, and our workman from New Zealand, Rod, was just wonderful. Tim was very good through that time too, but it wasn't fun for anyone in the family, to be honest.'

A generally fit and healthy person, Nancy has learnt in recent years that many of her health problems stemmed from her body being unable to metabolise certain drugs and chemicals, something she believes was triggered by not recovering properly from her first anaesthetic. Never one to focus on the negative, Nancy says her experience with Guillain-Barré, in particular, taught her patience. Even though the pain had gone, years later she was still unsteady on her feet at times and often dropped things. 'I was quite nervous about going up and down any sort of steps . . . and in the dark I feel like I'm walking like a duck. That's the most lasting thing, apart from the fact that since then I really haven't been able to stand late nights.'

It may be the start of winter, but in Casterton the sun is shining. It's the first day of the Australian Kelpie Muster and the wide main street is packed with people and dogs. Outside the red-brick Albion Hotel, spectators cram temporary stands to get the best view of the festival's most popular event—the kelpie high jump. Behind a bright yellow metal barricade, the handlers of competing dogs sit on rows of hay bales laid out on a strip of synthetic grass. At the far end of the strip is a white ute loaded with more hay bales. A simple metal frame placed behind its tailgate supports the jump itself—a series of pine planks stacked horizontally. As dogs progress through each round, more planks are added, until the solid face of the timber wall is metres high.

Clearing the jump requires teamwork. Each dog is usually accompanied by two people. One holds the dog tight a few strides back from the jump, while the other climbs onto the ute and leans over the wall to call the dog up and over. Some handlers resort to enticements to get their dog's attention and motivate them to scale the seemingly impossible obstacle, waving favourite toys or a meaty bone. In the early stages, most dogs make it easily, urged on by the barks of their rivals and enthusiastic cheers from the crowd. As the height increases, they have to dig in with their claws to scrabble up, making the most of their momentum to reach the top and hook their front paws over the highest plank. In 2016, a big, tan, three-year-old kelpie called Bailey set a new world record when he easily cleared 2.915 metres. Astonishingly, Bailey was a city interloper, a pet from suburban Melbourne who spent most of his day sleeping on his owners' bed and whined if he was left outside in the rain. He won the dash and the overall title too.

The second day of the festival has a more serious focus. In the afternoon, buyers gather for an auction that attracts international interest. Kelpies of various ages and stages of training go under the hammer. In the 25 years to 2022, a total of more than $3.2 million in sales was racked up, with a then record price for a single dog set in 2021 when a grazier from north-eastern Victoria paid $35,200 for an experienced dog named Hoover.

For a number of years, Nancy chaired the committee of the Casterton Kelpie Association responsible for organising the auction, offering her own dogs up for sale sometimes too. She has been part of this festival from the beginning, spending eleven years on the association executive and working alongside other members of the community to set it up in the mid-1990s. Like many other rural towns in those days, Casterton was struggling economically, with the population in decline and businesses and services fast disappearing. A working party reckoned it might be worth leveraging the area's claim as the birthplace of the kelpie to put Casterton on the map and attract more visitors.

The first visible manifestation of this dream was a bronze statue of a kelpie, created by award-winning Melbourne sculptor Peter Corlett. It was unveiled in the main street with great fanfare in 1996, to coincide with the town's 150th celebrations. Then Apex Club stalwart Ian 'Spud' O'Connell suggested the working-dog auction. Walking trails and more works of art followed, before a fully-fledged festival emerged in 2001, with a whole weekend of activities to draw in the crowds.

Nancy and local horseman Barry Murphy added to the program when they created a new event to honour the spirit of pioneering Australian stockmen and the first kelpies. In the Stockman's Challenge, a kelpie, rider and stockhorse work together to drive a small mob of sheep through a series of obstacles. The event is held under large river red gums where the Glenelg River flows through the town.

Today it's hard for anyone to miss the kelpie connection when they drive into Casterton. Taking pride of place in the main street is the Australian Kelpie Centre. The bold, modern, timber-clad building, which opened in 2018, features giant photographic portraits of kelpies on its external walls. Looking out across the street is one of Nancy's dogs, Sparrow, and near the front door is another—Anna. People are encouraged to pause in front of her and take 'a selfie with a kelpie'. Inside the centre, a carefully curated interpretative display tells the story of the breed. Deeply fascinated by its history, Nancy contributed to the research.

Like the Mitchells, the breed's bloodlines go back to Scotland. Kelpies are descended from dogs known as Scotch collies. Different from border collies, most came from the Highlands, where they had to cover huge distances in rugged terrain and extreme weather conditions, often working out of sight from their handlers. Their coats varied in length, but they were predominantly black and tan in colour, with very little, if any, white markings.

There are no written records of her exact pedigree, but the foundation female of the kelpie breed most likely descended from these Highland collies. Known as Gleeson's Kelpie she was whelped in around 1870 on Warrock station, about 30 kilometres north of Casterton. A black and tan bitch, she had a medium-length coat tinged with red, and semi-erect ears that came up when she worked. Kelpie caught the eye of Jack Gleeson, an itinerant stockman working on a neighbouring station, who clearly thought highly of her because he swapped a stockhorse to obtain the pup.

Jack later moved on to the Riverina where his friend, Mark Tully, managed a station in the Urana district. Tully gave him a small short-haired dog called Moss. Black with a smooth coat and erect ears, Moss had been bred near Yarrawonga on the banks of the River Murray by John Rutherford, whose working dogs were highly sought after. The son of a Highland shepherd, Rutherford imported dogs bred by his family for generations at Kildonan in the north of Scotland. An observer at the time noted that 'with Kelpie and Moss, Gleeson could do what he liked with sheep, whether few or many . . . both mustering and in the drafting yards.' The observer had never seen their equal.

Kelpie produced at least two litters with Moss. She was also mated with a dog called Caesar, whose parents were black and tan collies imported from Jedburgh—the birthplace of Nancy's Telford relations. According to Nancy, this line of dog threw the occasional red pup, and perhaps had more eye and style than other dogs. Caesar's sire, Brutus, wowed audiences in 1871 when he won one of the first sheep-dog trials held in Australia. Staged during the inaugural show at Burrangong near Young, the contest required dogs to gather up three sheep that had been let loose outside the showground and then bring them back onto the ground, through the crowd and into a pen. 'The performance of this dog was something wonderful,' wrote a correspondent to the *Sydney Mail*. 'So uncommonly well did this shepherd's friend behave himself that the other competitors resigned all claim to the prize, and would not put their dogs upon trial.'

One of the pups sired by Caesar was given to C.T.W. King, who named her Kelpie after her dam. Known as King's Kelpie, she also proved to be an outstanding trial dog. And the rest, as they say, is history, with kelpie finally catching on as the breed's name around the turn of the century. Today the working kelpie breed is described as a short-coated, prick-eared dog that revels in tough conditions and is capable of mustering huge areas and large mobs of sheep, with a highly developed ability to problem-solve.

Nancy has been living and breathing this heritage since she established the Pomanda Working Kelpie Stud in 1974. She thought a working-dog stud would fit in with station life, and give her something to do while potentially making money. Nancy was missing her work as a veterinary nurse, but the relative isolation of the station meant it wasn't feasible to take this up again. Continuing her nursing career was out too, because her wages would barely cover the cost of petrol.

Nancy's idea was to acquire a well-bred bitch and mate it with Tim's dog—a handy kelpie named Sam. She wasn't able to trace his bloodlines, but Sam came from an old strain of tall, lean, light-red dogs bred up the river near Morgan. Nancy has often wondered if he was descended from kelpies bred by local pastoralist and renowned working-dog authority, Gerald S. Kempe, who imported Rutherford dogs from Scotland in the late 1800s. 'Some of the paddocks at Nalpa were a thousand acres, with salt pan and samphire, and they weren't easy to muster. Sam was capable of mustering up to 800 sheep in a mob, and he was tough as nails in the shearing shed. He'd work all day,' says Nancy.

To make a start with her own breeding program, Nancy wrote to the Working Kelpie Council of Australia asking for advice about where she could source a well-bred bitch suited to station country. By good fortune, her inquiry was answered by Mike Donelan, owner of the well-known Bullenbong Kelpie Stud at The Rock, in southern New South Wales. He had a young pup that he thought would suit, but he was about to go droving so she would have to take it straight away.

Bullenbong Yarringa cost the equivalent of six weeks' pay for one stockman. She travelled by train all the way from Wagga Wagga. When Nancy collected her from the Murray Bridge railway station, she discovered a little pup with unusual colouring. 'We joked that her name was bigger than she was,' Nancy says. 'She sat up in my two hands, and she was light golden fawn and tan, with lovely black eyeliner, black tips around her ears, and a wide frieze of black hairs down her back and at the top of her tail. I'd never seen a kelpie like her.'

On the way home, Nancy stopped at the Nalpa cattle yards where Phil was working. 'It's a dingo! You've paid all that money for a dingo,' he exclaimed after one look at the pup.

Nicknamed Yarri, she grew up to be a beautiful dog and a good all-round worker, whose progeny were excellent. 'She was pure gold as a foundation bitch. You couldn't have found better,' says Nancy.

Over the next seven years, Nancy acquired several more dogs from Mike, visiting his property to meet him in person. Encouraging her to work with a kelpie's natural instincts, he told her that the trick with his mustering strain of kelpie was to let them run 'high, wide and handsome', leaving them to follow their instincts, while applying just a bare minimum of control.

Nancy also became fascinated by his definite ideas about breeding kelpies. Nancy didn't always agree, but she was prepared to listen. In line with a theory posed by Kempe, Mike believed the breed's ancestors came in two distinct types in terms of build and working style. 'One type of dog had an easygoing, loose slouchy style of work. They were tall dogs and very clever and they were the most highly valued of the Scotch shepherd dogs. They went out round the tops of hills mustering by themselves and you could trust them. They were bigger and had great courage and self-belief, and loved to stay with the sheep and look after them,' Nancy explains. 'Then you had the in-hand dog that has to be told what to do. They were generally more stylish dogs, with eye, but they required commands otherwise they could be rash and rush in to do things.'

Mike based his breeding program around these types, adamant that it was important to maintain pure strains of both. He was worried that the more independent type of kelpie was fast disappearing because it was less keenly sought after by farmers who mostly wanted general-purpose dogs that would follow commands and were easy to manage. So Mike had a family of more compact and forceful dogs that needed more input from the handler. They varied in colour and were more outgoing and energetic, without being excitable. Then he had a line of tall, mostly black and tan kelpies, that were calm and independent and worked wide. 'These are dogs you have to let go,' Nancy explains. 'You can't want to control them, because they don't really need you. If you try to control them with other than basic commands, they can be an absolute headache, but if you are prepared to let them think for themselves, then you've got the most special dog you could ever have.'

In 1982, Mike gave Nancy such a dog. Bullenbong Mate changed her understanding of what a kelpie is capable of doing, and made her a committed keeper of his centuries-old genetics. He came to Nancy after Mike sustained a severe back injury when he was thrown from a horse. Months later he was still recovering, so he rang her with a proposition. He had a seventeen-month-old kelpie that he had bred to be a sire. The dog had barely seen a sheep and was being wasted. Would Nancy like to take him on?

Bullenbong Mate was soon on his way by train. Accompanied by Simon, Jonathon and Chrissy, Nancy drove to the nearest railway station, at Bordertown. The platform seemed deserted, then she spotted a transport crate tucked around a corner, out of the sun and wind. 'I could see this black form in the crate so I walked over and I put my head down, and I said: "You must be Mate." And he looked at me, and that was it. No waggy tail, no nothing. The cage was locked so I turned to walk away and get somebody, and there was a woof, as if to say, "Hang on, I'm not staying here any longer." Well, a station attendant appeared with the key and out stepped this big handsome dog, with magnificent dark-chestnut markings. He gave a shake, looked around and then looked up at me. He was not fazed by anything.'

Nancy took Mate for a short walk then she put him in the back of her hatchback, and set off home with Chrissy and the boys. Dogs are not supposed to understand mirrors and how they work, but when Nancy looked into the rear-vision mirror to check how he was doing, Mate was looking back at her. 'You know, I think I'm going to like this dog,' she said. Before anyone had a chance to reply Mate gave one deep woof, and everyone burst out laughing.

Nancy could write a book about Mate and her experiences with this remarkable dog over the next seven years. Mike had promised her that working with Mate would teach her a great deal, and he was not wrong. 'I owe that dog so much,' reflects Nancy, looking up at a painting of him by talented Limestone Coast artist Jaime Prosser, which has pride of place above a fireplace in her home.

'Mate taught me fairly rapidly that he knew more than I did about livestock movement. I had to learn that I was not in control. You can't think that you are superior to a dog like Mate—you are not. You have to work on the same level,' she says. 'He taught me what dogs are capable of because basically he wouldn't listen. The first thing I noticed about him was that he would not sit, for love nor money. I settled for "stop" as the command to remain still, and he mostly obeyed. He taught me that if he couldn't trust me, I couldn't trust him. He taught me that if I sent him out of sight to look for stock, I had to remain where I was because if I moved he wouldn't know where I was.

'Mate would not kowtow to me, but he would have died for me. Everybody who worked with that dog was touched by him. They've never forgotten him. People said you had to see Mate working in the hills to have any idea what a dog could do. I often say to people that I feel like one of the apostles preaching a religion no-one else understands. Mike wrote to me in a letter that only one in a thousand people will understand the concept of allowing a dog to go off by itself to muster stock, with no control over what it's doing.'

When Nancy acquired Mate, she was living on Sugarloaf Hills, a 526-hectare property just down the road from Pomanda. Sugarloaf Hills was infested with rabbits and needed improvements, but she and Tim had snapped it up eighteen months after purchasing Pomanda because it came with a house. Pomanda didn't, so they had been living in a rented home near Kingston, and driving out every day.

Both properties were only partially developed for grazing, with large patches of natural scrub. Tim and Nancy worked hard to improve them, getting rid of the rabbits, sowing pasture and re-fencing to introduce a rotational grazing system that Tim had been keen to try for some time. Within a few years they built the stocking rates up to a thousand head of beef cattle and around seven thousand sheep, including a Suffolk stud they had started at Nalpa.

While Tim focused on tasks that involved working with machinery, Nancy took charge of stock husbandry. Because the country was rough and the flats prone to being inundated with water in spring, most of the stock work was done on horseback. It wasn't unusual for Nancy to spend four or five days a week on horseback, with the rotational grazing system requiring stock to be moved every week. She particularly enjoyed drafting cattle in the paddock, cutting out steers and then droving them to the stockyards at the far end of the property so they could be loaded onto a truck for market.

Although she had other horses, Nancy usually rode Breezy, who had grown into a pretty fine-boned mare with a flowing black mane. She would toss her head with excitement and take off across the paddocks when Nancy let her have her head, both revelling in the gallop as they moved from one watering point to another to check that everything was in order. 'I did some pretty wild riding in my life,' Nancy says, relishing the memory. 'Breezy carried me for thousands of kilometres over nine years at Pomanda and never once faltered. I was always carried safely despite the roughness of the country I rode her over, and the speed.'

Even Mate took to riding the mare. He learnt to jump up in front of the saddle and hitch a ride when they had long distances to cover between tasks. Perfectly balanced, he would rest his forepaws on her neck.

Initially, Nancy concentrated on developing Mate's instincts in the paddock. In one of her first experiences working with him, without being asked, the dog went after two freshly weaned lambs that had escaped from the yards. Mate rounded the first up while it was still in sight and brought it back to the yard fence, but the other disappeared 'like a rocket' into some ti-trees. Nancy expected the young dog to lose heart and abandon the task, but a few minutes later the lamb appeared, shepherded by Mate who blocked every attempt it made to bolt, keeping wide and applying just enough pressure when needed. 'You're a very good dog,' Nancy told him when the lamb was safely back in the yards.

The hills on Pomanda had red soil gullies between short, sharp limestone rises where sheep could disappear from view in seconds. One day, Nancy sent Mate after a mob of

400 wethers which had shot away behind a small rise. After waiting a few minutes with no sign of the sheep or dog, Nancy decided to walk behind the rise to check what was going on. There was no sign of them so she walked further. Still unable to find them, she retraced her steps and found Mate back in the original gully, holding the mob and occasionally standing up on his hind legs looking for Nancy, who was downwind. 'Needless to say, in future, when I cast him out of sight, I didn't move until he returned with the sheep.'

Mate was pretty handy in yards too. At about three o'clock in the morning on a cold winter's day, Nancy was woken up by the distant rumbling of a livestock transport pulling onto the property hours earlier than expected with a load of sheep. Heading out into the dark, she fetched Mate and drove to meet the truck. 'It's a girl!' the astonished driver greeted her.

He was even more astonished about what happened next. Working efficiently, with only an occasional whistle and command from Nancy, Mate quickly guided the sheep out of the truck and into the yards, while the driver huffed and puffed as he struggled to let the ramps down fast enough to keep up the flow. 'Good dog, lady!' he commented, amusing Nancy with the apparent graduation in his esteem, from girl to woman.

About a year after Nancy acquired Mate, she had the opportunity to prove the dog's abilities to a wider audience. A new type of competition for working dogs was emerging in Australia, in the form of yard trials. These trials test a dog's ability to negotiate sheep through a set of yards within a specified time limit, following a prescribed course that involves putting them through a drafting race and moving them up ramps, or other obstacles usually found in farm sheep yards. The concept was pioneered in Tasmania in the early 1980s, and first run on the mainland at the South East Field Days, held at Lucindale, about 70 kilometres from where Nancy lived. Shortly after this event, she answered a call for people interested in forming a South Australian Yard Dog Association. She attended the first meeting and remained an active member for many years, serving as president in the 1990s, and as the inaugural secretary of the Australian Yard Dog Association.

Nancy competed in trials for more than ten years, travelling to events in Victoria and New South Wales as well as in her home state. On several occasions, she was accompanied on longer drives by her mother. They travelled in Nancy's white V8 Holden ute, which was fitted with a specially designed mesh crate on the back, large enough to carry six dogs. To keep them warm and comfortable, Nancy stuffed bags full of straw or wool to create a soft bed, and covered the crate with a fitted canvas tarpaulin. Mate went on the early trips and did extremely well, being placed in many events, including state championships, but he often lost a crucial one or two points because he was too independent. 'Mate didn't listen. In his opinion I was there to get the gates!' Nancy jokes.

Nancy never won a state championship, but she was the first woman to judge an Australian Yard Dog Championship, and her dogs reached the finals on more than a few occasions. In 1989, she placed second in the Victorian Farm Dog Championship with Pomanda Iago, one of Mate's sons. However, his most outstanding direct progeny was Nacooma Gus, bred by Lyndon Cooper from his bitch Ollie. An outstanding trainer then living near Kingston, Lyndon described Gus as 'almost supernatural' because of his powerful presence and the way he worked sheep. The black and tan kelpie was placed more than 70 times and won at least 27 yard and farm trials, including five state championships.

While trialling was a great way to test her dogs, when it came to the Pomanda stud, Nancy realised it was more important to breed dogs that suited farmers, who are the mainstay of her business. 'It doesn't matter how good the dog is, if it's too hard to place with my clients, there is no point breeding that type,' she explains.

After mating Yarri with Mate, Nancy realised the pups were a little too independent in nature for most farmers, so she contacted John Gedye at Scoriochre stud near Dundonnell in western Victoria. He offered to mate Yarri for free with his dog, Scoriochre Gunga, whose mother was a Bullenbong dog. The progeny became the foundation of the Pomanda stud's A family.

Convinced that 85 per cent of a dog's abilities stem from genetics, Nancy's aim was to produce dogs with natural ability to look for sheep and then move them efficiently, saving time and reducing stress on both the livestock and the handler. Out in the paddock that meant going to the head of a mob so the dog could block and turn them effectively, and then 'balancing' the sheep so they moved forward rather than zig-zagging across a paddock. In the yards, she wanted them to have the right amount of force.

Experimenting with bloodlines and genetics takes time, and there are no guarantees pups will carry the desired traits. To test their potential, Nancy puts them on a lead when they are a few months old and walks them down the road past some paddocks with livestock. 'If a pup holds its head high and is looking, looking, looking, you know it's going to search for livestock,' she explains. Then she might take them into a hilly area, where the sheep are upwind and out of sight, watching to see if the pup's nose goes up. 'I look for all these little signals and dogs that look out and away from themselves. And I let them run high, wide and handsome when they are young. Some pups will run out ahead, some will stay near the bike, and some will stay naturally behind it. The ones that will muster are the ones that are running out there, because they are independent and sure of themselves. They require more thought to train, whereas the ones that are more dependent on you are easier to train.'

By the 1990s, Nancy was breeding around a hundred pups a year for clients spread across Australia. Many of her clients today are repeat customers, but with every inquiry she takes care to match the buyer with a dog best suited to their needs. While she breeds dogs to work in a variety of settings, Nancy realises some people are looking for a companion too. 'I do get photos of dogs on couches and in beds, and between the owners, which is a bit of a worry!' she laughs.

Many of her clients keep in touch, seeking advice or giving her updates on their dog's progress. And then there are the heartbreaking calls when Nancy becomes a grief counsellor, often struggling to set aside her own emotions so she can soothe distraught owners whose dogs have died unexpectedly.

Among them was a little red and tan bitch called Sorrell, a highly intelligent dog who used to sit alongside Nancy when she was sorting mail on the floor, and accompany her to the clothesline. Without any training, Sorrell started picking up and passing the envelopes and clothes to hang up. Nancy sold her to a repeat client named John. About twelve months later she started getting strange phone calls. Picking up the receiver all she could hear was heavy breathing. Nancy was starting to worry, when the person finally spoke. It was John. He had accidentally run over Sorrell and killed her. 'Every time I've rung to tell you, I've burst into tears and had to hang up. I walk to the back door and I still see her there. She just haunts me,' he confessed.

'What can you say? Obviously, I was a tad upset as well,' Nancy recalls sadly, quoting from a favourite Rudyard Kipling poem: 'You've given your heart to a dog to tear.'

All sorts of perils can take the life of a dog, but perhaps the one that upsets Nancy most is snakebite. Despite her best endeavours to keep them out of harm's way, she reckons she's lost something like fifteen dogs to venomous snakes over the years. And then there was the devastating impact of canine parvovirus when it emerged in Australia in the early 1980s. 'We had a dreadful time because dogs had no resistance and it wiped out pups like you wouldn't believe. I lost some of the best dogs I'd ever bred, a lot of them by Mate, and there was nothing you could do, even after hours of consultations with drug manufacturers and vets, because the new vaccines didn't work reliably.'

Then came the greatest loss of all. Her precious Mate died in March 1989. Only eight and a half, he wasn't old as dogs go. He'd been badly injured about eighteen months before when a horned sheep smashed into him in a raceway. He'd recovered from that, and then he'd started puffing and peeing blood. Nancy took him to her regular vet, who sent her to another practice in Mount Gambier so Mate could be x-rayed. Determined not to let him suffer, Nancy told them not to let the dog wake up from the required anaesthetic if it was as bad as she suspected. 'And that's what happened. He was absolutely full of cancer.'

Nancy was so devastated that she gave her trial dogs to other people to work. 'I could be driving down the road and just burst into tears, and that went on for months. I took him home and we buried him at the foot of this big blue gum tree. It was Tim, Jono and myself, bawling our eyes out.'

Just months before Mate died, the Withers had sold up and moved to a property at Lochaber. It wasn't by choice. In the mid-1980s, they'd sold Sugarloaf Hills and borrowed money to purchase the other half of Pomanda. Like hundreds of other Australians, on paid advice, they took advantage of the deregulated financial sector and borrowed overseas money through a hedging product offered by their bank. The idea was to take advantage of a strong Australian dollar to offset rising interest rates, but when the dollar crashed in 1985, borrowers were left with huge debts.

The stress for Tim and Nancy was enormous. 'It was a huge mistake. I am not trained in international finance and nor was my husband. The bank would ring us up and make us decide when to hedge, but we had no information to go on and it could all fall apart in an hour. We are not gamblers, but it was like playing the pokies, so we made up our mind to sell while we still had something left.'

The 485-hectare property at Lochaber came with some fine-wool merinos that they ended up crossing with White Suffolks to produce lambs for meat, after the wool market collapsed in the early 1990s. Nancy and Tim also maintained a healthy number of their own Simmental and shorthorn cattle.

To bring in more income soon after the move to Lochaber, Nancy took up an unexpected offer from her old employer, Ian McLachlan. In June 1989, he was preselected as the Liberal candidate for the safe federal seat of Barker. When he spotted Nancy at a local function organised to generate support, Ian asked if she would be interested in helping out during shearing at the family's flagship property north-west of Hay. Historic Tupra station ran 42,000 sheep. The manager needed someone to supervise the yards and help with sheep work for about six weeks, starting in mid-August. Then a call came asking Nancy to assist with shearing on another McLachlan property, Mena Murtee station, near Wilcannia. She suspects this shorter stint involving fewer sheep was conceived to test her capabilities before the Tupra job. Not keen on the idea despite the financial benefits, Tim agreed she could go.

Nancy had travelled as far as Wentworth with a ute-load of dogs when she received a phone call warning that shearing had been delayed because it was pouring with rain. After talking it over, Nancy met the station manager, Ian Botten, at the highway turnoff to the station so he could tow her vehicle along the 15 kilometres of dirt road to the homestead. 'What an experience! I was sideways, every which way, sliding all over this red dirt road,' she recalls.

The dogs were accommodated in a shed, while Nancy stayed in the homestead with Ian and his family. It was so wet for the first week, she rarely left the house apart from wading through clinging mud to feed her dogs and give them a run. In the end she was away three weeks. Tim was not happy when she finally got home. To mollify her husband, Nancy organised for her parents to come up and look after him while she was in the Riverina (by then Simon and Jonathon were both away at boarding school in Adelaide).

Nancy remains forever grateful that Ian McLachlan gave her the opportunity to work at Tupra, which turned into an annual gig that lasted fourteen years. 'What big landholder approached a woman in those days and offered her a job like that? He didn't have to do it, but there was a mutual respect,' she says. 'The best part was that I would take six trained dogs and about the same number of young dogs. I could test those dogs in the most extreme conditions, and work out which were the best ones, which were the ones I wanted to keep and breed from, and which ones I'd sell.'

Nancy's kelpies were kept on chains in the station's old stables, while she stayed with the manager, Chris McClelland, who was also an accomplished artist, and his wife, Margie, a talented photographer. Nancy enjoyed their company and hospitality. They took care of all her washing, and she didn't have to do any cooking or cleaning. Nancy appreciated this rare luxury immensely, especially given she was putting in twelve-hour days in the yards.

Nancy's job during shearing was mammoth. So was the Tupra woolshed, which reminded her of an aircraft carrier because of its imposing mass. Built in the early 1900s on the banks of the Lachlan River, it originally had 50 stands, with the Australian Workers' Union newspaper (*The Worker*) reporting in 1913 that 132,800 sheep had been pushed through in seven and a half weeks. In Nancy's time the property and sheep numbers were much smaller, but there might be up to fifteen shearers, averaging about 2600 sheep a day between them. The biggest daily tally was 3200 merino hoggets.

It was Nancy's responsibility to make sure the shearers had a steady supply of sheep. She usually worked on her own with four dogs to fill the shed each day, before shearing started—a sizeable task given there was room for 2600 sheep inside the shed and another 800 underneath. She also had to count the shorn animals out, and oversee their drenching, branding and drafting. Free of their fleeces, sometimes the big merino wethers would jump out of the pen and take off into a huge yard that might be holding as many as two thousand sheep. One of Nancy's best kelpies, Rio, would fly over the fence after each miscreant, single it out and bring just that sheep back to the fence, where he would hold the animal until a jackaroo grabbed it.

Nancy was also responsible for supervising staff working in the yards, where she had the assistance of at least one jackaroo. Mostly young men, they didn't always know what to make of her when they first met, given there were very few women working in a similar capacity at the time. Her diminutive height didn't help. In her first year on Tupra, she worked mainly with Dave, also in his first year on the station. A full 30 centimetres (a whole foot!) taller than Nancy, he was a laconic lad with a lazy voice and a great sense of humour.

Within a few days of her arrival, Dave decided to put Nancy to the test. 'This particular day we were shearing lambs, and one of them was too sick to stand up, so I dragged it out and I said to Dave, "We'll have to put it down. It's too ill." And Dave straightened up to his full height and handed me his pocket knife.

'What Dave didn't know was that I'd been responsible for killing all the sheep for meat for my family for years, and I'd killed I don't know how many for the dogs too. It was a fairly big pocket knife so I cut the lamb's throat very professionally. I knew it was a test and I was never asked to do that again at Tupra. The young jackaroos were told, "This bird's okay", but every now and then there would be a casual helper, male of course. Some of them could be difficult but I never felt I had to prove myself to anybody, not ever—it was my choice whether I did or not—because of the foundation I was given by my parents growing up. They gave me self-belief in my ability to do things.'

That confidence came in handy in 1996, when Nancy was chosen as the South Australian Rural Woman of the Year. The nomination started as a joke. Nancy was in a back paddock, wielding a large knife and an axe, when Jonathon and his then girlfriend came out to see her. She was cutting up a dead cow to provide meat for the dogs. 'I think we might nominate you for rural woman of the year,' Jonathon jested, taking in the bloody scene.

Winning the award opened doors for Nancy, who received leadership and corporate governance training as part of her prize. Over the coming years she was invited to serve on a range of boards and advisory authorities for federal and state governments. Two of them were committees representing rural women, formed to advise the then Australian minister for primary industries John Anderson. She also served as president of the Foundation for Australian Agricultural Women, set up in 1995 to represent rural women and advance their involvement in agricultural occupations and rural communities.

These organisations aside, there were numerous times over the years when Nancy was one of maybe only two women on a state board or regional committee, at a time when women were vastly under-represented in the rural sector. 'Often, I was a "token" woman, however, I'm a strong believer in the token woman because if you're not there you don't have any influence whatsoever. However you got there really doesn't matter, as long as it's not underhand and you are capable of doing the job,' she says.

One of the experiences Nancy particularly enjoyed was being a founding member of a new organisation in the Limestone Coast region formed to support women who had created their own businesses, or wanted to set one up. The aim of Women in Business and Regional Development was to provide them with access to skills and mentoring that might help them succeed. Many were entrepreneurial farm women, striving to generate alternative streams of income to sustain families after the wool industry crashed. 'I have to say it was one of the best things that I've ever been involved in. It's still going and it has made a massive difference to a lot of people.'

Nancy also developed a specific interest in vocational training. It began in the mid-1980s when TAFE asked her to run a course at Millicent for people wanting to know how to train working dogs. Later on, she gained formal qualifications in training and assessing and ran intensive week-long courses at other locations across the state, as far north as Leigh Creek. Another regular fixture was spending a few days every year teaching students at Longerenong Agricultural College near Horsham, in the Victorian Wimmera. The experience motivated her to write a small book, *Your Kelpie*, as a basic introduction to the kelpie breed and working-dog training, which has been reprinted numerous times. Under a new national scheme, she also became one of the first trainers to teach farm safety.

For about nine years, Nancy also chaired a federally-funded regional program encouraging vocational education in secondary schools. 'In those days a lot of schools were very resistant to it, and it was a bit difficult at times, but I believed so strongly in it. All we had in mind was to offer students who didn't want to pursue a university career a way to stay in school longer. If they wanted to be a carpenter, they could work for a carpenter one day a week and it would count towards their secondary qualifications. If they found they didn't like it, they hadn't lost anything. I think we achieved a lot, but then there was a change of government and the program ceased.'

After years of being unhappy in the relationship, Nancy left Tim in 2003 with the support of her two sons, and started building a new life managing a farm at Sandford, a few kilometres south-east of Casterton. Even though her former husband passed away in 2019, she is reluctant to go into details, except to say that the stressful times they had been through took a toll on the marriage. She had tried to remain positive and had worked hard to hold the family and farm together, but Tim had handled the pressures very differently.

Nancy first visited the 243-hectare Sandford property that became her new home as a favour to former neighbour and friend, Louise Johnson. She wanted Nancy's opinion on whether it would be a sensible investment. Even though it was rundown, it was good grazing country and Nancy thought the price being asked represented value for money. 'Right, I'm going to buy it and I want you to manage it for me,' Louise proposed.

Nancy cites this as another example of the great opportunities she has been given in life by other people. It also reflects her willingness to grab them with both hands. 'Opportunity can be very fleeting and you have to grasp it,' she says.

She left Lochaber with little more than her clothes, a couple of recipe books, some art gear, and a green Mazda ute she had purchased from Jonathon, fitted with a crate for the dogs. She only brought a few dogs with her because there was nowhere to house them. The rest were left in the care of Lisa, a former Longerenong student whom Nancy paid to live onsite at Lochaber, until she could build some kennels.

Her friend turned 'boss' left it up to Nancy to buy suitable livestock based on an agreed budget. She started with a thousand crossbred ewes and later set up a breeding herd of mostly Murray grey cattle, crossed with Charolais and some Simmentals. They produced weaners that thrived on pasture and found ready buyers. Nancy was also allowed to run a certain number of cattle of her own, which she bought in a joint venture with Jonathon whenever she spotted a bargain at the saleyards.

Over the years, Nancy re-fenced the farm, realigning paddocks and gates in places to make them more practical, and installed a new watering system to supply the livestock and house. A workman came on Tuesdays to help with tasks that required two people, but otherwise she mostly managed on her own, apart from getting in contractors for seasonal work such as shearing and baling hay.

Nancy quickly came to love the farm, its rolling hills, towering gums and stunning views out across country that was the traditional home of the Gunditjmara people before the Henty brothers showed up from Tasmania in the 1830s and claimed most of it for their expanding flocks. Renamed Braeside, the farm was part of Sandford station before it was subdivided for closer settlement in the late 1800s and purchased by the Humphries family. Their original farmhouse is long gone now, but a scattering of surviving fruit trees brings whimsical sprays of blossom to the paddocks in spring.

A talented amateur artist, Nancy drew inspiration from the smooth-trunked sugar gums and majestic river red gums that still grace the property. Her favourite was a magnificent sprawling red gum she dubbed the 'guardian of the spring' because it grew near a freshwater spring that fed into a small creek. The cattle and sheep loved to shelter under the tree's shade in warmer months; and she often spotted black cockatoos in its branches.

Nancy was also drawn to the bare, windy hilltop overlooking the lower part of the property. On a clear day she could see the Grampians, purple-grey in the far distance. 'You could be quite close to nature there. If a storm came in from the west and you were up there checking on lambing ewes, it could get interesting. Sometimes I thought I might blow away! But I can't tell you how beautiful it was when the sky was blue with clouds forming, and there was light and shade running over the hills. In the summer it was all golden and beautiful, and in the winter it was green. When I came out really early in the morning in summertime and the sun was just coming up, the light was glorious. Everywhere I looked I could see a painting.'

Much to Nancy's dismay, commercial blue gum forests crept into the local area while she was there. Giant chippers ran day and night when the first generation of trees was ready for harvest, and the trucks carrying their output regularly kept her awake at night as they rumbled along the main road down in the valley.

It was one of the reasons Nancy was not sad to go when Louise decided to sell up in 2019. Rural land prices were booming and it was time to capitalise on the investment. Given she was almost 70, Nancy decided to retire from farming but not breeding and training kelpies. She found a house on a nearby property, and took the dogs with her. Among them is Titan, the patriarch of her A family, now thirteen and the dog closest in working type to Mate that she has ever bred. Five generations of his descendants came too, including his daughters, Theia, Tora, Quest and Lasca, and numerous granddaughters and grandsons. Nancy helps move sheep on the property, providing plenty of opportunity for the dogs to work, and still owns some cattle which are kept on agistment.

These days the Lochaber property is in the hands of Simon and his wife, Rachael, an agricultural scientist. After leaving school, Simon notched up valuable experience working on interstate livestock and cropping enterprises, and drill-stem testing with a cousin's business in the oil and gas industry. He spent five years as a field manager with a cotton enterprise before coming back to Lochaber, where he and Rachael are raising two children, Eliza and Lachlan. Meanwhile, Jonathon studied rural business management at Glenormiston agricultural college in western Victoria, and worked for a range of rural supply companies. He also took on drill-stem testing before joining a business that specialises in building livestock-handling facilities. He lives a few minutes away from Nancy on a small farm with his wife, Nicola, a veterinary surgeon who ran her own practice in Casterton for several years, and their three children, Liam, Zoe and Alexandra.

Nancy is no longer on the kelpie festival organising committee because she doesn't enjoy driving at night to attend meetings, but after too many years to recall she remains a Victorian representative on the Working Kelpie Council of Australia, which meets monthly by phone. 'It's very worthwhile,' she says, recounting her experiences tackling the Victorian government over proposed legislation intended to stop unscrupulous dog breeders from operating puppy farms. The unwitting consequence was that it would have forced working-dog studs like Nancy's to shut down.

After representing the council at a meeting where city bureaucrats seemed completely uninterested in anything her sector had to say, she harnessed her experience on government boards to help raise the alarm and demand proper consultation. 'The livestock working dog has contributed an enormous amount to the Australian economy; there are many things we could not have achieved without them, and they'd not even spoken to us,' she says with lingering frustration.

Eventually a parliamentary inquiry was called. Its findings deplored the lack of proper consultation, and the government withdrew the bill. After 'years of hell to get there', the government passed revised legislation in 2017 which exempted farm working dogs. In the meantime, Nancy became a founding member of the Australian Federation for Livestock Working Dogs—formed during these years to make sure governments across the country had access to reliable information before they made decisions that affected working dogs and their husbandry. Nancy devoted hundreds of hours to doing research and helping with submissions as other states tackled similar legislation. The lack of recognition given to livestock working dogs, even by farmer lobby groups, still infuriates her, especially given their contribution to the Australian economy every year, which the federation has calculated at something like a billion dollars.

Nancy's determination to make this case reflects her ongoing passion for kelpies. It also speaks to her wider love of animals and the respect she believes they are due. Having seen what Mate and his descendants are capable of doing, she thinks it is arrogant to believe that man ranks above dogs, or any other animal for that matter, in the hierarchy of life. She attempts to explain with a favourite quote from American naturalist Henry Beston:

> We need another and a wiser and perhaps a more mystical concept of animals . . . In a world older and more complete than ours they move finished and complete, gifted with extensions of the senses we have lost or never attained, living by voices we shall never hear. They are not brethren,

> they are not underlings; they are other nations, caught with ourselves in the net of life and time, fellow prisoners of the splendour and travail of the earth.

The quote comes to mind when Nancy visits Lochaber and stands under the tree sheltering Bullenbong Mate's grave. Other much-admired dogs are now buried there too. 'What would I give to have you all back again?' she ponders wistfully.

Acknowledgements

Researching and writing *A Farming Life* has been a drawn-out process, more so than almost any other book I have written so far. I was part way through the research phase when a certain pandemic made it impossible to continue, since my approach involves a considerable amount of travel and staying with the women on their properties so that I can absorb something of their daily lives, listen to their stories and meet their families.

After discussions with the publisher, the book was put on hold indefinitely, waiting for the world to right itself. Who knew it would take almost two years before the last interstate travel restrictions were removed in Australia? By then life had changed for one of the women I had already visited and she reluctantly withdrew, while another had moved farms. Then, when the manuscript was only weeks away from being completed, a crisis within my own family made it difficult for me to write for some time.

I would like to thank Allen & Unwin for standing by the book and by me during these unforeseen delays and challenges.

In particular, my heartfelt thanks to commissioning editor Tessa Feggans and in-house editor Courtney Lick, for their endless patience and tender care nursing me and the book through its final stages. I am so proud of the end result.

Thanks also to my agent Fiona Inglis, for sticking with me, and the band of 'Moggies' who have provided continual support and encouragement since our little group of professional writers first came together fifteen years ago.

Usually, I find myself apologising to friends and family who are often badly neglected while I hide myself away to write. After the pandemic forced the whole world into isolation, they might not even have noticed this time!

But I do want to thank my home community in the Adelaide Hills, which has made me feel very welcome since I moved here five years ago. At the height of lockdowns, complete strangers volunteered to collect my mail, walk the dogs and shop for groceries. They even shared precious supplies of toilet paper.

As always, love and gratitude go to my family for their ongoing support. Your love and the solid grounding my parents gave me will always lie at the very heart of my own resilience.

And last but not least to Amber, Kelly, Ruth, Helen, Kristen, Donna, Belinda, Michelle and Nancy. Thank you for trusting me to tell your stories, and for your open-hearted approach to sharing your lives and the land you love. It is an honour and a privilege that I don't take for granted.